Drywall

Level 2

SECOND EDITION

P **Pearson** **NCCER** | National Center for Construction Education and Research

NCCER

President and Chief Executive Officer: Boyd Worsham
Vice President of Innovation and Advancement: Jennifer Wilkerson
Chief Learning Officer: Lisa Strite
Senior Curriculum Leader: Chris Wilson
Senior Manager, Production: Graham Hack
Drywall Project Manager: Patrick Bruce
Technical Writing Manager: Gary Ferguson
Technical Writer: Jo Ann Bartusik
Art Manager: Bree Rodriguez
Production Artist: Judd Ivines, Chris Kersten
Permissions Specialists: Adam Black, Sherry Davis
Project Coordinators: Colleen Duffy, Milagro Maradiaga

Pearson

Commercial Product Manager – Associations: Andy Dunaway
Senior Digital Content Producer: Shannon Stanton
Associate Project Manager: Monica Perez-Kim
Content Producer: Alexandrina Wolf
Executive Marketing Manager: Mark Marsden
Designer: Mary Siener
Rights and Permissions: Jenell Forschler
Composition: Integra Software Services
Printer/Binder: Lakeside Book Company
Cover Printer: Lakeside Book Company
Text Typefaces: Palatino LT Pro and Helvetica Neue

Cover Image
Cover image of three workers finishing and sanding drywall next to a window. Cover photo provided by Marek Brothers LLC.

1 2024

National Center for Construction Education and Research

ISBN-10: 0-13-821061-6
ISBN-13: 978-0-13-821061-8

PREFACE

To the Trainee

Walk into almost any home, apartment complex, or commercial building and look around. The odds are good that drywall applicators installed the walls and ceilings and placed insulation, soundproofing, and firestopping materials behind and onto those walls and ceilings. They may also have applied textures and trims to enhance both the interior and exterior of the buildings.

There were approximately 97,000 drywall and ceiling tile installers working in the United States in 2021. Careers can progress from installer to specialty finisher to business owner, and related professions include sheetrock applicator and acoustical carpenter.

In the first level of drywall training, you learned the basics, including insulation, installation, and finishing. In this second level, you will explore specialized topics, such as steel framing, acoustical ceilings, and specialty finishes. Both levels provide you with information and knowledge about the materials and tools used in the drywall profession.

By choosing to pursue drywall training, you are taking a step toward a satisfying and rewarding career. Continuing your craft education is important, as technology and the materials are changing all the time. We wish you success as you begin your construction career in the drywall trade, and hope that you will continue your training outside of this series. By taking advantage of training opportunities as they arise, you will demonstrate initiative and a desire to learn—qualities that are present in the industry's best professionals.

New with Drywall Level 2

The revised second edition of Drywall Level Two has been modernized to reflect current tools and practices of the craft. NCCER is proud to release this edition with our latest instructional systems design, linking learning objectives to each module's content. This revised edition expands content around cold-formed steel framing applications to establish more balance between residential and commercial drywall applications.

We wish you success as you progress through this training program. If you have any comments on how NCCER might improve upon this textbook, please complete the User Update form using the QR code in the frontmatter of this book. NCCER appreciates and welcomes its customers' feedback. You may submit yours by emailing **support@nccer.org**. When doing so, please identify feedback on this title by listing *#DrywallL2* in the subject line.

Our website, **www.nccer.org**, has information on the latest product releases and training.

NCCER Standardized Curricula

NCCER is a not-for-profit 501(c)(3) education foundation established in 1996 by the world's largest and most progressive construction companies and national construction associations. It was founded to address the severe workforce shortage facing the industry and to develop a standardized training process and curricula. Today, NCCER is supported by hundreds of leading construction and maintenance companies, manufacturers, and national associations. The NCCER Standardized Curricula was developed by NCCER in partnership with Pearson, the world's largest educational publisher.

Some features of the NCCER Standardized Curricula are as follows:
- An industry-proven record of success
- Curricula developed by the industry, for the industry
- National standardization providing portability of learned job skills and educational credits
- Compliance with the Office of Apprenticeship requirements for related classroom training (*CFR 29:29*)
- Well-illustrated, up-to-date, and practical information

NCCER also maintains the NCCER Registry, which provides transcripts, certificates, and wallet cards to individuals who have successfully completed a level of training within a craft in NCCER's Curricula. *Training programs must be delivered by an NCCER Accredited Training Sponsor in order to receive these credentials.*

For information on NCCER's credentials and the NCCER Registry, contact NCCER Customer Service at 1-888-622-3720 or visit **https://www.nccer.org**.

Digital Credentials

Show off your industry-recognized credentials online with NCCER's digital credentials!

NCCER is now providing online credentials. Transform your knowledge, skills, and achievements into digital credentials that you can share across social media platforms, send to your network, and add to your resume. For more information, visit **www.nccer.org**.

Cover Image

At MAREK, they believe a company is like a building: it is only as strong as the foundation on which it is built. Their foundation is forged from an 86-year commitment to quality and service unequaled by anyone in the industry. In 1938, MAREK entered the drywall industry and almost immediately changed it with the way they purchased, delivered, and installed drywall in the market. Through their strong industry relationships and partnerships, they have since diversified their portfolio to include ceilings, acoustical treatment, fabric panels, flooring, paint, shades, and day lighting solutions as well as turn-key remodeling, and maintenance and improvement services in major markets extending from locations in Texas, Tennessee, and Georgia.

As MAREK has grown, so has their responsibility as a role model in the industry and in the communities they serve. They are an active leader in industry organizations and strive to instill their ethics and values. MAREK's industry leadership also includes technology innovations like Building Information Modeling (BIM) to deliver faster, safer, and more efficient projects with general contractors, architects, and owners. For over 80 years, MAREK has shaped the industry by constantly striving for excellence.

Today, they continue to raise their work standards through innovations like BIM, prefabrication, workforce development initiatives like the Construction Career Collaborative (C3), and much more. In every aspect of their work, they exceed expectations and remain true to the values and principles that have always defined the company. This allows MAREK to give their customers piece of mind as they work on their most recent projects as well as future projects. That is the MAREK difference – **strength from within.**

DESIGN FEATURES

Content is organized and presented in a functional structure that allows trainees to access the information where they need it.

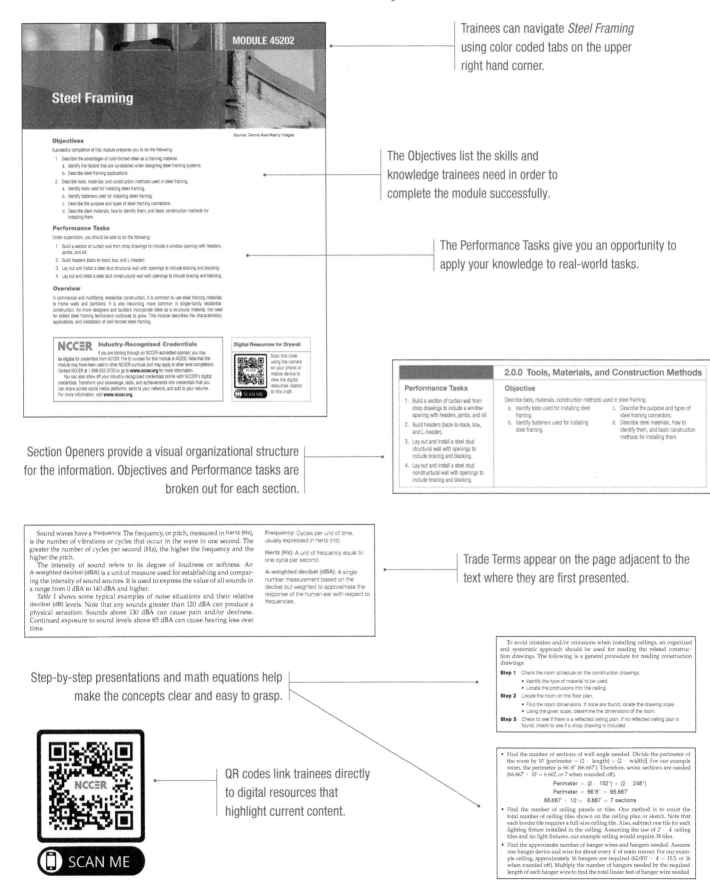

Trainees can navigate *Steel Framing* using color coded tabs on the upper right hand corner.

The Objectives list the skills and knowledge trainees need in order to complete the module successfully.

The Performance Tasks give you an opportunity to apply your knowledge to real-world tasks.

Section Openers provide a visual organizational structure for the information. Objectives and Performance tasks are broken out for each section.

Trade Terms appear on the page adjacent to the text where they are first presented.

Step-by-step presentations and math equations help make the concepts clear and easy to grasp.

QR codes link trainees directly to digital resources that highlight current content.

Important information is highlighted, illustrated, and presented to facilitate learning.

Placement of images near the text description and details such as callouts and labels help trainees absorb information.

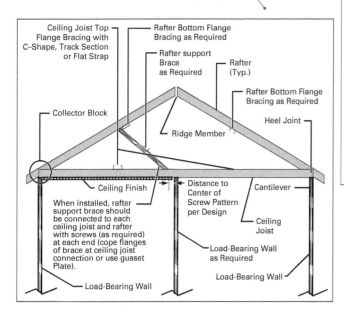

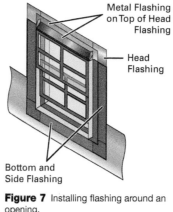

Figure 7 Installing flashing around an opening.

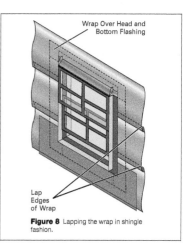

Figure 8 Lapping the wrap in shingle fashion.

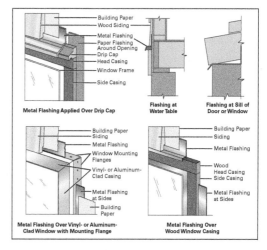

WARNING!

Do not drill or install powder-actuated fasteners into any type of post-tensioned concrete because you might damage the tendon, which may fly out of the slab, causing injury. Damaged tendons also reduce the load-bearing capacity of the beam or slab.

CAUTION

Be careful when removing ceiling material from its packaging and when handling it. Keep all ceiling material and the grid system clean and undamaged.

NOTE

Make sure all inspections for areas and/or equipment above the grid line have been completed before beginning the ceiling installation.

New boxes highlight safety and other important information for trainees. Warning boxes stress potentially dangerous situations, while Caution boxes alert trainees to dangers that may cause damage to equipment. Notes boxes provide additional information on a topic.

Going Green

Digital Plans

The days of printing and carrying large stacks of paper plans are over. Although paper versions are still preferred by some, and even necessary in certain situations, the convenience and efficiency of digital plans is taking over. Construction professionals have their choice of software that can be used on a tablet, laptop, or even on a smartphone to view plans and manage other project-related documents from any location. It is inevitable that some revisions to the plans will become necessary during a large commercial project. The process of documenting these revisions and noting them in the revision block of the plans is known as *drawing revision control*. Drawing revision control ensures that everyone involved in the project is working from the most up-to-date set of plans. The use of construction management software makes the most up-to-date document sets available to everyone as soon as they are uploaded into the program. This allows work to continue without delay and saves a few trees in the process.

Going Green looks at ways to preserve the environment save energy, and make good choices regarding the health of the planet.

Think About It

Fiberglass Insulation

1. Are the flanges on faced fiberglass insulation always stapled to the inside of the stud?
2. Can you increase the effectiveness of fiberglass insulation by squeezing more into a smaller space?

Hold-Down Clips

Some ceiling tiles require hold-down clips to secure the ceiling tiles to the grid. For example, clips are used with lightweight tiles to prevent them from reacting to drafts. One manufacturer specifies that clips be used if the tiles weigh less than 1 pound. Hold-down clips are not necessarily required for ceilings used in fire-resistance-rated applications. Check the manufacturer's instructions.

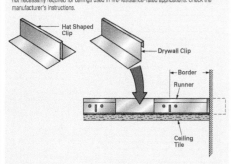

These boxed features provide additional information that enhances the text.

Did You Know?

Steel Framing Is Customizable

Prefabricated metal framing systems can be customized to create unique design elements. All or part of a structure's frame can be engineered to specifications and assembled off-site. Customized studs, wall panels and systems, clips and fasteners, floor joists, and roof trusses can be ordered and installed as separate pieces on-site. Customization of steel framing materials gives designers the ability to create eye-catching, contemporary architecture while taking advantage of the strength and durability of steel.

Review questions at the end of each section and module allow trainees to measure their progress

---

1.0.0 Section Review

1. The main difference between building paper and building wrap is that the wrap creates a nearly airtight structure whereas the paper is water_____.
   a. proof
   b. resistant
   c. permeable
   d. tight

2. Most building wrap is made of spun, high-density_____.
   a. polyethylene fibers
   b. glass fibers
   c. cement fibers
   d. paper fibers

3. True or False? A different method of installing building wrap should be used if windows and doors have already been installed.
   a. True
   b. False

Module 45205 Review Questions

1. The primary purpose of exterior cladding is to _____.
   a. hide blemishes
   b. protect the structure
   c. look good
   d. provide insulation

2. When wrapping a building, the beginning of a new roll should overlap the end of the previous roll by _____.
   a. 1" to 6"
   b. 6" to 12"
   c. 12" to 18"
   d. 18" to 24"

3. What is the purpose of a control joint?
   a. Drain water from behind the cladding
   b. Cover vertical seams to prevent water infiltration
   c. Support the insulated panels of EIFS
   d. Relieve coating stress due to thermal changes

ACKNOWLEDGMENTS

This curriculum was revised as a result of the farsightedness and leadership of the following sponsors:

Association of the Wall & Ceiling Industry
Baker Triangle
DPR Construction
FCI Cumberland
Grayhawk, LLC
Marek Brothers

This curriculum would not exist were it not for the dedication and unselfish energy of the volunteers who served on the Authoring team. A sincere thanks is extended to the following:

Bill Ford
Johnny Hull
Kevin Howser
Ricardo Menchaca
Ricardo Reyes Aguilar
Richard Bell
Don Allen

NCCER PARTNERS

To see a full list of NCCER Partners, please visit
www.nccer.org/about-us/partners.

CONTENTS

Module 45201 Commercial Drawings

Module 45202 Steel Framing

Module 45203 Suspended and Acoustical Ceilings

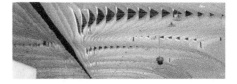

Module 45204 Interior Specialties

Module 45205 Exterior Cladding

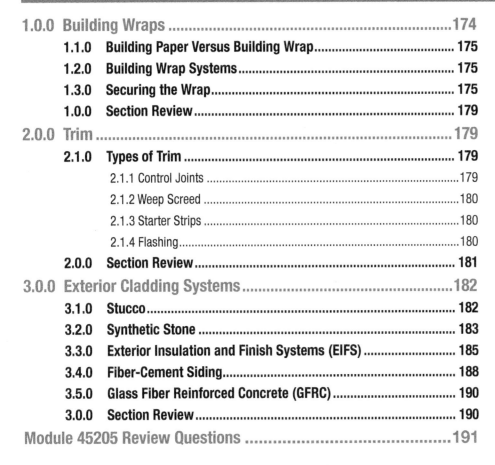

Module 45206 Interior Finishes

Commercial Drawings

Source: Noah Clayton/Tetra Images/Alamy Images

Objectives

Successful completion of this module prepares you to do the following:

1. Explain the difference between commercial and residential drawings.
   a. List the components of a commercial drawing set.
2. Identify drawings and information contained in a set of plans.
   a. Identify architectural drawings and describe the information found in them.
   b. Identify structural drawings and describe the information found in them.
   c. Identify mechanical, electrical, and plumbing (MEP) drawings and describe the information found in them.
   d. Explain the role of specifications and describe the information found in them.

Performance Tasks

Under supervision, you should be able to do the following:

1. Find the following components within a commercial drawing set: reflected ceiling plan, wall sections/details, structural framing plan, door schedules, wall schedules, MEP plans, and specifications.
2. Locate the information in a commercial project's specifications that apply to the installation and finishing of drywall.
3. Locate a door in a commercial drawing. Using a door schedule, identify the characteristics of the door that should be installed, including the type, size, material, hardware, rating, and finishing.
4. Locate a window in a commercial drawing. Using a window schedule, identify the characteristics of the window that should be installed, including the type, size, material, hardware, rating, and finishing.

CODE NOTE

Codes vary among jurisdictions. Because of the variations in code, consult the applicable code whenever regulations are in question. Referring to an incorrect set of codes can cause as much trouble as failing to reference codes altogether. Obtain, review, and familiarize yourself with your local adopted code.

Overview

Your role as a drywall technician may include installing, framing, finishing, insulating, soundproofing, and firestopping walls and ceilings. Regardless of your contribution to a project, skilled workers from other trades who follow behind you will depend on your work to be done accurately and exactly according to plan. For this reason, you will need to be skilled in reading and understanding commercial drawings. The basic principles of scaling, dimensioning, measuring, and calculating are fundamental for interpreting both residential and commercial drawings. However, commercial drawings are much more detailed, and can contain additional structural, mechanical, and electrical systems. Knowing how to identify and read the different types of commercial construction plans as well as the notes, schedules, and specifications that supplement the plans, is an important skill for drywall framing technicians. Commercial plans require close study and review. Your understanding of them will help keep construction activities well-organized and true to design.

Digital Resources for Drywall

Scan this code using the camera on your phone or mobile device to view the digital resources related to this craft.

SCAN ME

1.0.0 Commercial Drawings

Performance Task

1. Find the following components within a commercial drawing set: reflected ceiling plan, wall sections/details, structural framing plan, door schedules, wall schedules, MEP plans, and specifications.

Objective

Explain the difference between commercial and residential drawings.
   a. List the components of a commercial drawing set.

A large part of the construction market in the United States is commercial and industrial construction. Commercial buildings come in various shapes and sizes (*Figure 1*). To build commercial structures, you must be able to understand and interpret an architect's plans and drawings. Commercial plans and drawings are usually more complex than residential plans. This module will

Figure 1 Examples of modern commercial buildings.
Source: (Top) swalls/Getty Images; (Bottom) Peter_Fleming/Shutterstock

introduce you to commercial drawings. The basic principles are the same as those that apply to residential drawings. However, commercial structures have more dimensions to interpret, additional section and detail drawings, and more changes in scale.

Commercial or industrial construction work is often called *heavy construction*. This is because heavy equipment and heavy materials are used for these jobs. Cranes, bulldozers, graders, and excavators are just a few examples of heavy equipment. Heavy materials include steel and concrete.

Most commercial projects are larger and more complicated than residential projects. They require a greater variety of construction techniques, equipment, and materials.

Consequently, plans and drawings for a commercial project are also more complicated than residential plans and drawings. Commercial structures have many different uses. Safety and environmental requirements must also be considered. The complexity of the commercial project is reflected in the complexity of the plans.

Commercial construction plans are more detailed than residential plans for several reasons. First, the structures are usually larger and more expensive to build. Another major consideration is legal liability. The contractor's legal liability is far greater in commercial construction because there are more applicable codes, ordinances, and regulations. Other reasons commercial plans are more complex include the following:

- The architectural plans, drawings, **schedules**, and specifications are legal documents. Many state, local, and federal agencies demand greater detail in commercial construction drawings to substantiate any legal disputes.

- Code restrictions and safety requirements for commercial and industrial buildings are far more complicated than those for residential construction. More detailed drawings are used to make certain that all codes and local ordinances are met.

- The size of a commercial building requires a greater number of drawings, **sections**, details, and schedules, with more detail required to correlate the various parts of the structure.

- The materials used in commercial construction call for more detailed information on construction techniques, especially for structural steel.

A major commercial project may have hundreds of drawings in its plan set, plus associated schedules. Specialty subcontractors will generally get a partial set of plans relating to their work. However, they should request a full set, including the room finish schedules. A variety of electronic applications are available for use on mobile devices for viewing and managing plans in the field. When paper plans are being used, it is a good idea to keep at least one complete set in the field office for reference.

During the course of a project, **as-built** drawings are created using a set of working drawings, sometimes called *record drawings* or *redline drawings*. As-built drawings incorporate (through revisions or annotations) any changes of dimensions, materials, form, and method of construction encountered in the completion of the structure or site.

All changes, no matter how minor, as well as all clarifications are entered into a change database and recorded by the general contractor on the working drawings. Some examples of changes and clarifications are *engineering change proposals* (ECPs), *change orders,* and *requests for information* (RFIs). ECPs are changes that are communicated by the manufacturer to address a problem or improve the function or design of a product. Change orders are revisions that are made to the scope of work in the contract after a project has begun to deal with an unexpected problem. All changes and modifications go through a formal approval process involving the general contractor, subcontractors, engineers, and project managers.

Once the project is complete, the as-builts are formalized into a permanent set of as-built drawings by the architectural firm and copies are distributed to the project owners as well as other designated parties.

Schedules: Charts or tables that provide detailed information corresponding to various parts of the drawings.

Sections: Drawings that detail construction techniques or materials.

As-built: Revised drawing of an installation that shows the changes made during installation.

NOTE

While some builders are sticking with the traditional method of redlining drawings with pen and paper to record changes for as-builts, construction management software has made it possible to redline drawings from tablets and laptops at the jobsite. Regardless of which method is used, it is critical that all changes are properly noted and included in the as-built drawings.

Going Green

Brick Upon Brick

The Empire State Building was completed in 1931 and for nearly 40 years it was the tallest building in the world. Approximately 10 million bricks were used in its construction. Today it is likely that curtain walls containing brick facades would be used in place of individual bricks.

Despite the enormity of the project, construction of the Empire State Building was completed in about 15 months. One of the methods used to speed up construction was to have trucks unload the bricks down a chute, instead of dumping them in the street. The chute led to a large hopper from which the bricks were loaded into carts and hoisted to the location where they were needed. This innovative technique eliminated the back-breaking work of using a wheelbarrow to move bricks from the brick pile to the bricklayer.

Even though the Empire State Building was built in the 1930s, it has undergone extensive upgrades and is a charter member of the ENERGY STAR® building label program. ENERGY STAR® is a joint program of the US Environmental Protection Agency and the US Department of Energy that helps to protect the environment by using energy-efficient products and practices.

1.1.0 Components of a Drawing Set

Construction drawings consist of several different kinds of drawings assembled into a set (*Figure 2*). You may notice that the components of a commercial plan set are very similar to those that you would see in residential construction. Of course, commercial plan sets will be much larger and more complex. A complete set of commercial construction plans typically includes the following drawing types: civil/site plans, architectural plans (floor plans, interior/exterior elevations, sections, details, and schedules), structural drawings (foundation plans, framing plans), and MEP plans (mechanical, electrical, and plumbing). Each type of drawing is assigned a letter. Architectural plans are labeled with the letter *A*, structural plans are *S*, and so on. For each type, there may be several drawings, which are then numbered. For example, the first few electrical drawings would be numbered E1, E2, and E3.

The exact content and organization of the plan set will vary, depending on the size of the job, local code requirements, and the preferences of the design firm. For example, a drawing set for a commercial office building may include **landscape drawings**. The letter *L* would identify these. In some commercial drawing sets, the site plans are called **civil drawings**. They are marked C1, C2, C3, and so forth. Specifications and schedules may be referenced in a drawing and included separately or included with each type of drawing. A helpful first step in reading a set of plans is to familiarize yourself with the index of drawings (*Figure 2*) which shows the exact organizational structure used by the designer on that project.

Landscape drawings: Drawings that show proposed plantings and other landscape features.

Civil drawings: Drawings that show the overall shape of the building site. They are also called *site plans*.

INDEX TO DRAWINGS :

ARCHITECTURAL
T-1.0	TITLE SHEET
LS-1.0	LIFE SAFETY PLAN
SP-1.0	SITE PLAN
A-1.0	FLOOR PLAN
A-1.1	CANOPY PLAN & DETAILS
A-1.2	REFLECTED CEILING PLAN & DETAILS
A-1.3	ENLARGED ROTUNDA PLAN & RESTROOM PLANS
A-1.4	ENLARGED COLUMN PLAN DETAILS
A-1.5	ROOF PLAN
A-2.0	BUILDING ELEVATIONS
A-2.1	BUILDING ELEVATIONS
A-3.0	BUILDING SECTIONS
A-3.1	WALL SECTIONS, CANOPY SECTION & DETAILS
A-4.0	SECTION DETAILS
A-4.1	SECTION DETAILS
A-4.2	CABINETRY PLANS & SECTION DETAILS
A-5.0	DOOR & WINDOW DETAILS
A-5.1	DOOR & WINDOW DETAILS
A-6.0	FINISH PLAN, SCHEDULES & DETAILS
A-6.1	INTERIOR ELEVATIONS
A-6.2	INTERIOR ELEVATIONS
A-6.3	INTERIOR ELEVATIONS
A-6.4	FURNITURE PLAN

STRUCTURAL
S-1.0	STRUCTURAL NOTES, SCHEDULES & DETAILS
S-2.0	FOUNDATION PLAN & NOTES
S-2.1	MAIN LEVEL ROOF FRAMING PLAN & DETAILS
S-2.2	UPPER LEVEL ROOF FRAMING PLAN & DETAILS
S-2.3	CANOPY ROOF FRAMING PLAN & DETAILS
S-3.0	FOUNDATION DETAILS

MECHANICAL
M-1.0	MECHANICAL PLAN & NOTES
M-2.0	MECHANICAL SCHEDULES & DETAILS

ELECTRICAL
E-1.1	ELECTRICAL SITE PLAN & DETAILS
E-2.1	LIGHTING PLANS & NOTES
E-3.1	POWER PLANS & NOTES
E-4.1	ELECTRICAL DETAILS & SPECIFICATIONS
E-5.1	ELECTRICAL RISER & DETAILS, SCHEDULES

PLUMBING
P-1.1	PLUMBING FOUNDATION PLAN & NOTES
P-2.1	PLUMBING PLAN & NOTES
P-3.1	PLUMBING SCHEDULED & DETAILS

TECHNOLOGY
TE-1.1	TECHNOLOGY PLANS, DETAILS & NOTES
TE-2.1	TECHNOLOGY DETAILS
TE-2.2	TECHNOLOGY DETAILS

Figure 2 Typical index for a commercial drawing set.
Source: Photo courtesy of Sanders & Associates Architectural Services

1.1.1 Drawing Elements

Title block: A section of an engineering drawing blocked off for pertinent information, such as the title, drawing number, date, scale, material, draftsperson, and tolerances.

Although there will be some variation in the exact elements used in each drawing set, most drawings consist of the same basic parts: a **title block**, *border*, *drawing area*, *revision block*, and *legend*. The title block is normally located in the lower right-hand corner or across the right edge of the drawing and contains basic information such as the name of the design company, the project name, the name of the draftsperson, issue and revision dates, and the drawing number. The revision block is usually found near the title block and is used to record any changes to the drawing in sequential order, including a brief description of the change, the date, and the initials of the person who made the revision. *Figure 3* is an example of a title block. *Figure 4* shows a basic revision block.

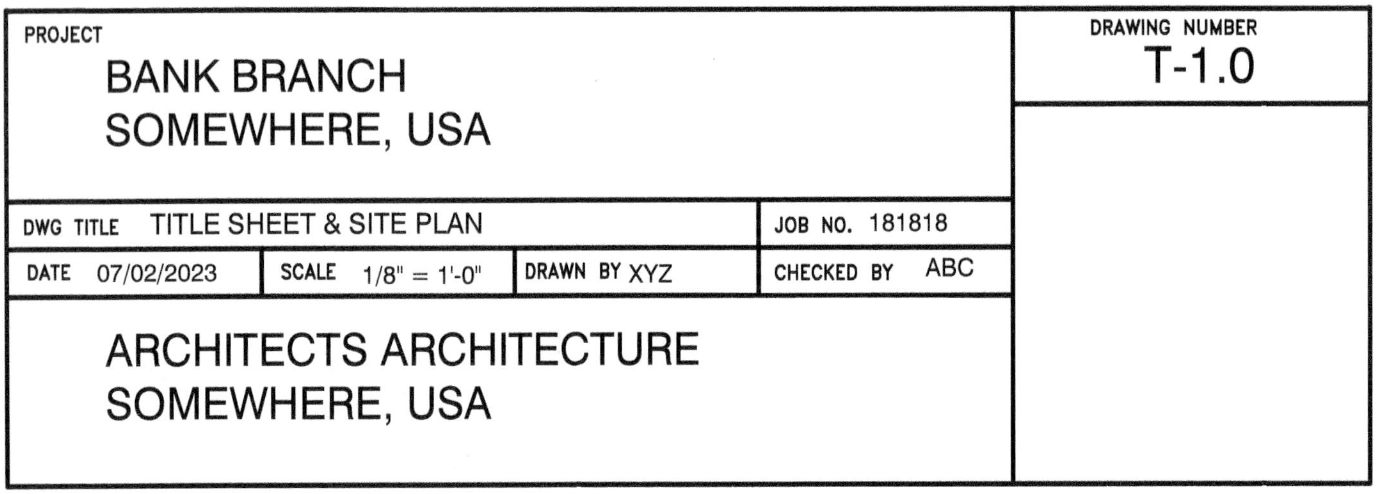

PROJECT			DRAWING NUMBER
BANK BRANCH **SOMEWHERE, USA**			**T-1.0**
DWG TITLE TITLE SHEET & SITE PLAN		JOB NO. 181818	
DATE 07/02/2023	SCALE 1/8" = 1'-0"	DRAWN BY XYZ	CHECKED BY ABC	
ARCHITECTS ARCHITECTURE **SOMEWHERE, USA**			

Figure 3 Title block.
Source: Photo courtesy of Sanders & Associates Architectural Services

REVISIONS		
No.	Description	Date
	FOR STATE APPROVAL	07/03/23
	BID SET	07/23/23
1	ADDENDA 1	07/30/23
2	ADDENDA 2	08/13/23
3	CONSTRUCTION SET	10/22/23
4	CLARIFICATIONS	12/06/23
5	BANK EQUIPMENT	12/19/23
6	CLARIFICATIONS	01/22/24
7	PORCELAIN TILE CHANGE - DUE TO SCHEDULE	04/24/24

Figure 4 Revision block.
Source: Photo courtesy of Sanders & Associates Architectural Services

Within a commercial drawing set, a variety of drawing elements—symbols, abbreviations, and drawing lines—are used. The most common drawing lines are the following:

- *Dimension line* — Establishes the dimensions (sizes) of parts of a structure. These lines end with arrows (open or closed), dots, or slashes at a termination line drawn perpendicular to the dimension line.
- *Leader and arrowhead* — Identify the location of a specific part of the drawing. They are used with words, abbreviations, symbols, or keynotes.

- *Property line* — Indicates land boundaries.
- *Cut line* — Identifies part of a drawing that is to be shown in a separate cross-sectional view.
- *Section cut* — Shows areas not included in the cutting line view.
- *Break line* — Shows where an object has been broken off to save space on the drawing.
- *Hidden line* — Identifies part of a structure that is not visible on the drawing. You may have to look at another drawing to see the part referred to by the lines.
- *Centerline* — Shows the measured center of an object, such as a column or fixture.
- *Object line* — Identifies the object of primary interest or the closest object.

Figure 5 shows common drawing lines, and *Figure 6* is a sample of a legend of symbols for a set of commercial drawings.

Symbols provide another graphic indication for different types of materials. Although there are many frequently used symbols, there are no standardized symbols for specific materials. Most engineers and architects use symbols that have been adopted by the American Institute of Architects (AIA) and the American National Standards Institute (ANSI). However, many firms use their own modified symbols to suit their specific needs. For this reason, drawings typically have a legend or key that lists the symbols and abbreviations used throughout the plan set. *Table 1* is an example of some of the many abbreviations used in commercial plans.

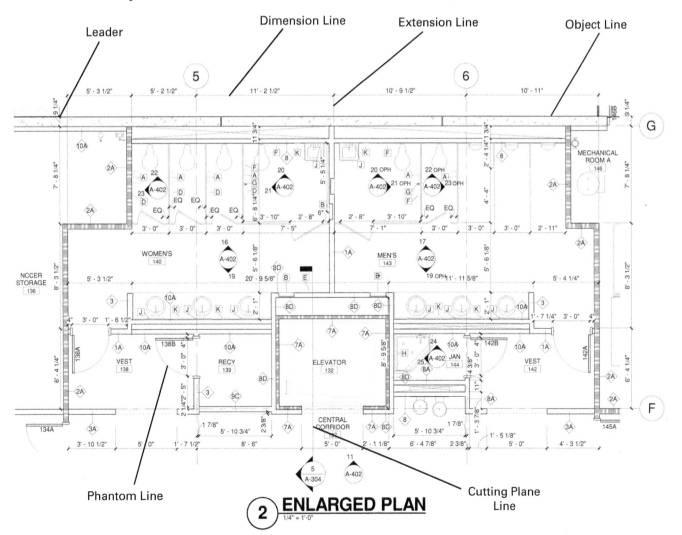

Figure 5 Common drawing lines.

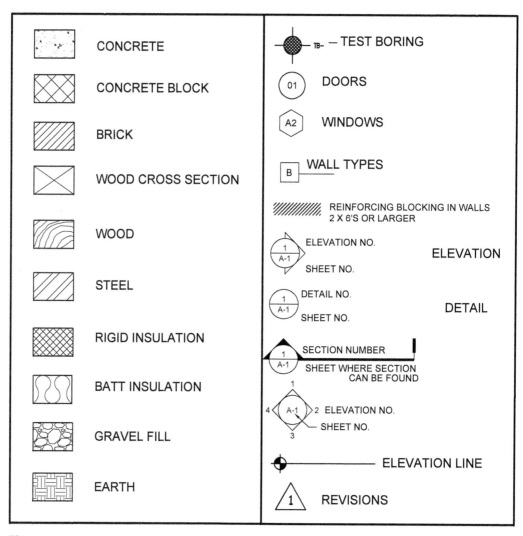

Figure 6 Example legend of symbols.

Source: Photo courtesy of Sanders & Associates Architectural Services

As you become more skilled at reading commercial drawings, you will begin to remember the standard abbreviations, symbols, and drawing lines. Fortunately, memorizing this vast amount of information is unnecessary unless you want to impress your fellow workers. Most commercial drawing sets will

Source: sculpies/Getty Images

Going Green

Digital Plans

The days of printing and carrying large stacks of paper plans are over. Although paper versions are still preferred by some, and even necessary in certain situations, the convenience and efficiency of digital plans is taking over. Construction professionals have their choice of software that can be used on a tablet, laptop, or even on a smartphone to view plans and manage other project-related documents from any location. It is inevitable that some revisions to the plans will become necessary during a large commercial project. The process of documenting these revisions and noting them in the revision block of the plans is known as *drawing revision control*. Drawing revision control ensures that everyone involved in the project is working from the most up-to-date set of plans. The use of construction management software makes the most up-to-date document sets available to everyone as soon as they are uploaded into the program. This allows work to continue without delay and saves a few trees in the process.

TABLE 1 Common Abbreviations Used in Commercial Drawings

Abbreviation	Meaning	Abbreviation	Meaning
A.B.	Anchor bolt	GA.	Gauge
ADJ.	Adjacent	GR.	Grade
A.F.F.	Above finished floor	H.S.B.	High strength bolt
A.I.S.C.	American Institute of Steel Construction	INT.	Interior
ALT.	Alternate	JNT.	Joint
ARCH.	Architectural	MECH.	Mechanical
A.S.T.M.	American Society for Testing and Materials	MAX.	Maximum
BLDG.	Building	MIN.	Minimum
BM.	Beam	N.I.C.	Not in contract
B.O.	Bottom of	NOM	Nominal
CANT.	Cantilever	N.T.S.	Not to scale
CH.	Chamfer	O.C.	On center
C.J.	Control/construction joint	R.	Radius
COL.	Column	REQ'D.	Required
CTRD.	Centered	SCHED.	Schedule
DIM.	Dimension	SIM.	Similar
DWG.	Drawing	SPEC.	Specification
EL.	Elevation	SQ.	Square
EQ.	Equal	STD.	Standard
EXT.	Exterior	SYM.	Symmetrical
FDN.	Foundation	U.N.O.	Unless noted otherwise
FIN.	Finish	VAR.	Varies
F.O.W.	Face of wall	V.I.F.	Verify in field
FTG.	Footing		

include a glossary of abbreviations. In addition, a wide variety of reference tools are available online and in print format to look up abbreviations and symbols used in construction plans.

1.1.2 Architectural Drawings

Architectural drawings are usually labeled with page numbers beginning with the letter *A*. These contain general design features of the building, room layouts, construction details, and materials requirements. The architectural drawings will include the following:

- A site plan and/or other location plans
- Floor plans
- Wall sections
- Reflected ceiling plans
- Door and window details and schedules
- Partition details (*Figure 7A* and *Figure 7B*)
- Elevations
- Special application details including finish details and schedules

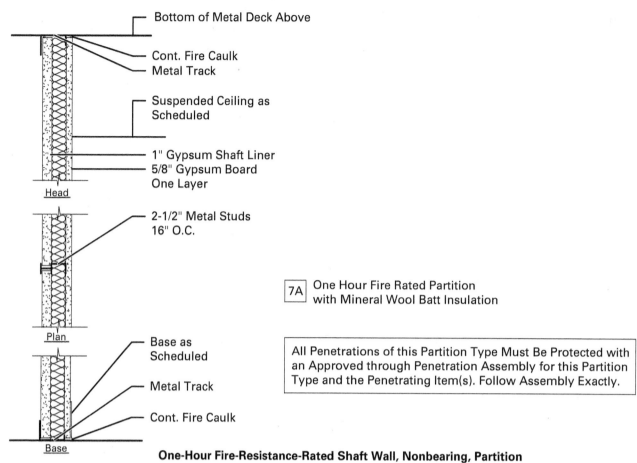

Bottom of Metal Deck Above

Cont. Fire Caulk
Metal Track

Suspended Ceiling as Scheduled

1" Gypsum Shaft Liner
5/8" Gypsum Board One Layer

Head

2-1/2" Metal Studs 16" O.C.

Plan

| 7A | One Hour Fire Rated Partition with Mineral Wool Batt Insulation |

All Penetrations of this Partition Type Must Be Protected with an Approved through Penetration Assembly for this Partition Type and the Penetrating Item(s). Follow Assembly Exactly.

Base as Scheduled

Metal Track

Cont. Fire Caulk

Base

One-Hour Fire-Resistance-Rated Shaft Wall, Nonbearing, Partition

Figure 7A Partition details (1 of 2).

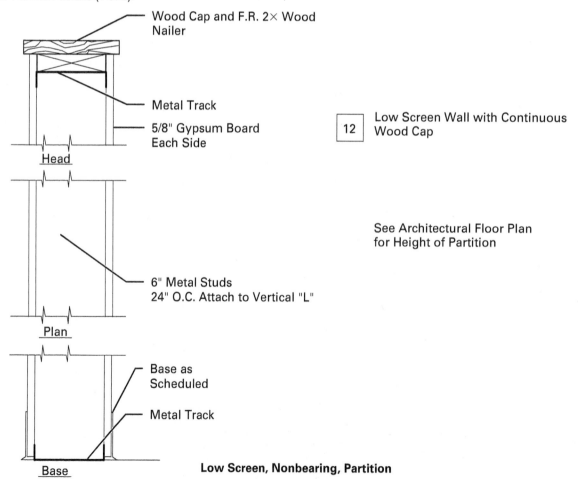

Wood Cap and F.R. 2× Wood Nailer

Metal Track

5/8" Gypsum Board Each Side

Head

| 12 | Low Screen Wall with Continuous Wood Cap |

See Architectural Floor Plan for Height of Partition

6" Metal Studs 24" O.C. Attach to Vertical "L"

Plan

Base as Scheduled

Metal Track

Base

Low Screen, Nonbearing, Partition

Figure 7B Partition details (2 of 2).

The site plan shows topographic features including trees, bodies of water, and ground cover. It will also show infrastructure such as roads, railroad tracks, and utility lines. Typical topographic symbols are shown in *Figure 8.*

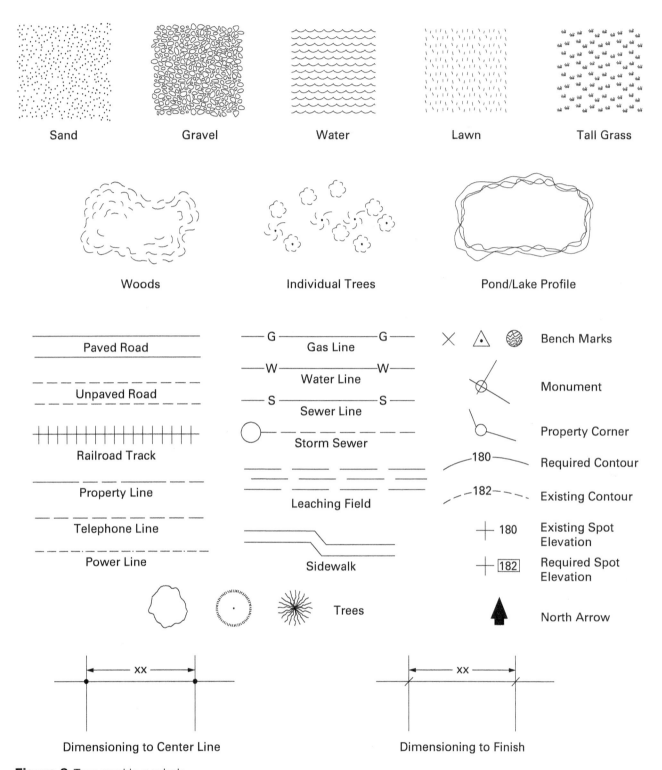

Figure 8 Topographic symbols.

1.1.3 Structural Drawings

Structural drawings provide a view of the structural members of the building and how they will support and transmit those loads to the ground. Structural drawings are numbered sequentially and designated by the letter S. They are normally located after the architectural drawings in a plan set.

Structural engineers prepare structural drawings. They must calculate the forces on the building and the load that each structural member must withstand. The structural support information includes the foundation, size, and reinforcing requirements; the structural frame type and size of each member; and details on all connections required. The structural drawings usually include the following:

- Foundation plans
- Structural framing plans for floors and roofing
- Structural support details
- Notes to describe construction and code requirements

Structural support for a commercial building may be steel framing, precast concrete structural elements, or cast-in-place concrete. Unlike residences, modern commercial buildings do not have wooden frames, except in unusual circumstances.

Structural drawings provide useful information. They can stand alone for craftworkers such as framers and erectors. The structural drawings show the main building members and how they relate to the interior and exterior finishes. They do not include information that is unnecessary at the structural stage of construction.

Structural drawings start with the foundation plans. Foundation plans are followed by ground-floor or first-floor plans, upper-floor plans, and the roof plan. Only information essential to the structural systems is shown. For example, a second-floor structural plan would show the steel or concrete framing and the configuration and spacing of the load-bearing members. The walls, ceilings, or floors would not be shown.

1.1.4 Mechanical, Electrical, and Plumbing (MEP) Drawings

MEP drawings show the different mechanical, HVACR (heating, ventilation, air conditioning, and refrigeration), electrical, plumbing, and fire protection systems of the building. You will often hear the terms "HVACR" and "mechanical" used interchangeably when referring to plan sets. Although most of a mechanical plan set will be the HVACR system, any system with mechanical components may be included. The **plan view** format is commonly used on these drawings. This view offers the best illustration of the location and configuration of the work. The drawings serve as a diagram of the system layout.

A large amount of information is required for mechanical work. There is limited space on the drawing to show the piping, valves, and connections. Special symbols are used for clarity. *Figure 9* contains some common HVACR symbols.

Detail drawings are sometimes used on mechanical drawings. Unlike the details shown on architectural drawings, these detail drawings are not normally drawn to scale. Usually, they are drawn as an elevation or perspective view. They show details about the configuration of the equipment.

Plan view: A drawing that represents a view looking down on an object.

Plumbing Drawings

Plumbing drawings are considered part of the mechanical plans. However, they are usually placed in their own set of drawings for clarity. Unless the building is very basic, placing them on the same sheets as the HVACR drawings would cause confusion. Plumbing drawings usually include the following:

- Site plan for water supply and sewage disposal systems
- Floor plans for the fire system, including hydrant connections and sprinkler systems
- Floor plans for the water supply system and fixture location
- Floor plans for the waste disposal system
- **Riser diagrams** to describe the vertical piping features
- Floor plans and riser diagrams for the gas lines

Riser diagrams: Isometric drawings that depict the layout, components, and connections of a piping system.

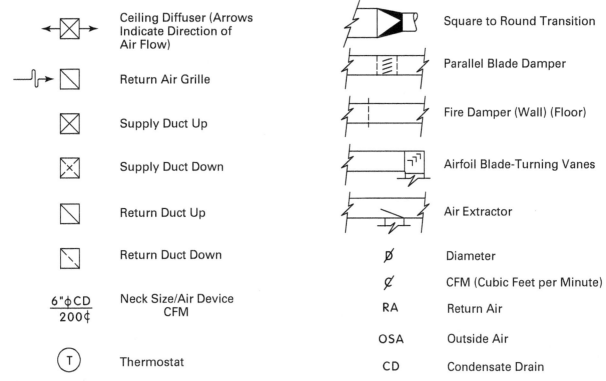

Figure 9 HVACR symbols.

The most common symbols used to designate the water and gas systems on plumbing drawings are shown in *Figure 10*. In addition to these symbols, there will be specific graphic symbols for the layout of such items as toilets, sinks, water heaters, and sump pumps. These will vary from structure to structure, depending on the specific design.

Electrical Drawings

The last part of the plan set usually contains the electrical drawings. They show the various electrical and communications systems of the building.

The electrical drawings are labeled with page numbers beginning with the letter *E*. They contain information on the electrical service requirements for the building. They also show the location of all outlets, switches, and fixtures. In addition to schematics of the branch circuits for the building, electrical drawings will include the following:

- Site plan for electrical service requirements
- Floor plans for the outlet and switch locations and the branch circuit requirements
- Lighting plans
- Emergency power and lighting systems
- Life safety systems
- Any backup power generation facilities
- Notes and details to describe other parts of the electrical system

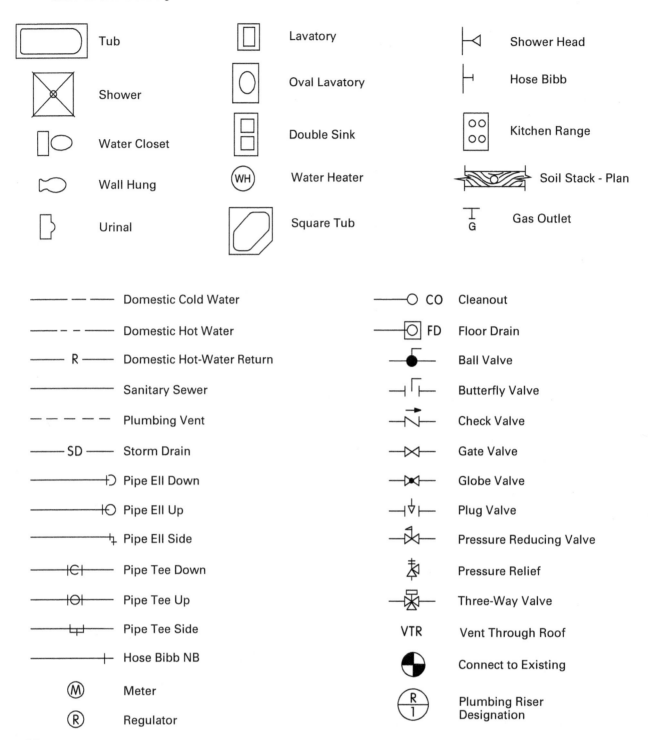

Figure 10 Plumbing symbols.

Like mechanical drawings, electrical drawings use the plan view to show system layout. Details and schedules provide clarification. One drawing may include power, lighting, and telecommunications layouts. In more complex structures, the systems are shown separately. There are many different symbols for electrical connections and fixtures. Commonly used symbols are shown in *Figure 11*. Special symbols for components such as power supplies, security systems, and circuit boards are usually designated by the manufacturer.

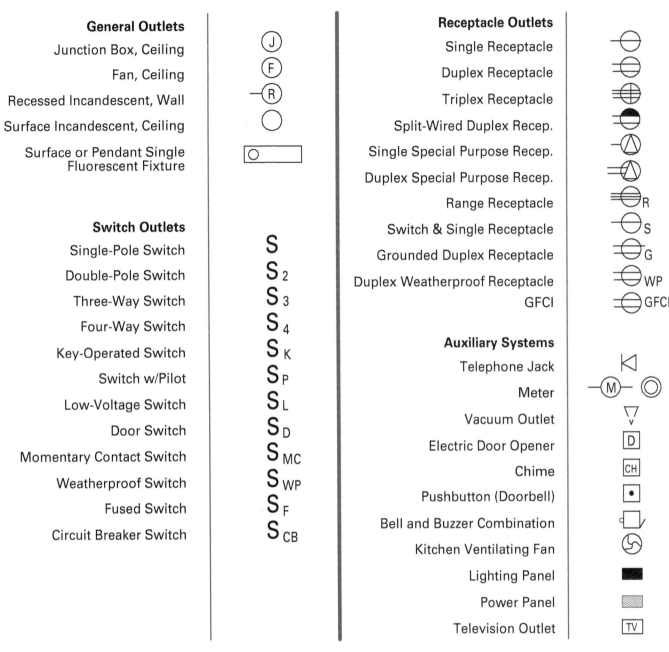

Figure 11 Electrical symbols.

1.1.5 ADA Requirements

All trades should be aware of the **ADA requirements** for other trades and coordinate whenever possible. For example, if you are finishing a bathroom and note that the plumbers put the toilet trap at the wrong floor spacing, it is best to note this before finishing and avoid a callback. Common ADA requirements include such things as doorway width and hardware height, plumbing fixture height, electrical switch and receptacle height, thermostat location, and many more.

ADA Requirements: The Americans with Disabilities Act requires the use of accessible design methods for the construction of public accommodations such as lodging, public parks, schools, and restaurants.

1.0.0 Section Review

1. How are commercial plans different than residential plans?
   a. Commercial plans contain different types of drawings, but residential plans only have architectural drawings.
   b. Residential plans contain architectural drawings instead of structural drawings.
   c. Residential plans are usually more complex.
   d. Commercial plans are usually more complex.

2. What is another name for civil drawings?
   a. Site plans
   b. Commercial plans
   c. Layout drawings
   d. Elevation views

3. The main categories of commercial drawings are architectural, structural, and _____.
   a. mechanical, electrical, and plumbing (MEP)
   b. landscape
   c. CAD
   d. schematic

4. Which part of a plan set is used to record any changes to the drawing in sequential order including a brief description of the change, the date, and the initials of the person who made the change?
   a. Callouts
   b. Revision block
   c. As-builts
   d. Change order

2.0.0 Reading Drawings and Specifications

Performance Tasks

2. Locate the information in a commercial project's specifications that apply to the installation and finishing of drywall.

3. Locate a door in a commercial drawing. Using a door schedule, identify the characteristics of the door that should be installed, including the type, size, material, hardware, rating, and finishing.

4. Locate a window in a commercial drawing. Using a window schedule, identify the characteristics of the window that should be installed, including the type, size, material, hardware, rating, and finishing.

Objective

Identify drawings and information contained in a set of plans.

a. Identify architectural drawings and describe the information found in them.

b. Identify structural drawings and describe the information found in them.

c. Identify mechanical, electrical, and plumbing (MEP) drawings and describe the information found in them.

d. Explain the role of specifications and describe the information found in them.

When reading commercial plans, it is best to follow a step-by-step process to avoid confusion and to catch all the important details. Use the following list as a guide:

Step 1   Begin by reading the project specifications to pick up details not found on the drawings. Remember that in cases of conflicting information, the specifications take precedence over the drawings, and the structural drawings take precedence over the architectural drawings.

Step 2 Quickly review all drawings that give a general impression of the shape, size, and appearance of the structure, including the site plan, floor plans, and exterior elevation views.

Step 3 Begin correlating the floor plans with the exterior elevations, making certain that the parts appear to fit together logically.

Step 4 Next, look at the wall sections and determine the wall types (materials, load-bearing, or nonbearing) and construction details or procedures.

Step 5 Review the structural plans. Determine the foundation requirements and the type of structural system. Turn back to the site plan, floor plans, wall sections, and elevations as often as needed to correlate the structural and architectural drawings.

Step 6 Review all details on the architectural and structural drawings. Carefully consider all items that may require special construction procedures. It may be necessary to label detail drawings with related floor plan and elevation information. This is known as creating a reverse trail from the detail drawings back to the larger view.

Step 7 Review all interior elevations and try to get a clear picture of what the interior of the building will look like.

Step 8 Review the finish schedule.

Step 9 Review the mechanical and electrical plans. Coordinate with other trades to ensure proper sequencing.

If any part of a drawing is not clear, check with your supervisor. There may be a logical explanation for the item in question, or the plans may require clarification. An RFI is generated to resolve these conflicts. The architect/engineer should clarify and resolve all conflicts in writing so there will not be any confusion later in the project.

> **Elevation views:** Drawings giving a view from the front, rear, or side of a structure.

NOTE

A **request for information (RFI)** is used to clarify any discrepancies in the plans. If there is a discrepancy, notify your supervisor. The supervisor will write up an RFI, explaining the problem as specifically as possible and putting the date and time on it. The RFI is submitted to the superintendent, who passes it to the general contractor, who passes it to the architect or engineer, who then resolves the discrepancy. Always refer to specifications and the RFI when deciding how to interpret drawings.

> **Request for information (RFI):** A document used during the construction process that is used to clarify and resolve questions and information gaps in plans, drawings, specifications, and agreements.

Drawing Revisions

When a set of drawings has been revised, always make certain that the most up-to-date set is used for all future work. Digital versions of drawings can be updated in real time so that everyone is working with the revised set of plans. If paper copies are being used, either destroy the old, obsolete drawing or else clearly mark on the affected sheets: Obsolete Drawing—Do Not Use. A good practice is to remove the obsolete drawings from the set and file them as history copies for future reference.

Also, when working with a set of construction drawings and written specifications for the first time, thoroughly check each page to see if any revisions or modifications have been made to the original. Doing so can save time and expense for all concerned.

CAUTION

If paper plans are being used instead of digital plans, always keep two sets (one in the office, and one in the field). Never remove original drawings or sheets from the set of original drawings in the office. Make sure that both sets are updated with any as-builts and other revisions or change orders. Pay attention to the revision date and applicable color codes to ensure that the correct drawing set is being used.

2.1.0 Architectural Drawings

Architectural drawings are the core drawings of any plan set. They are sequentially numbered, usually starting with the site plan or the basement. In some cases, the exterior drawings, such as the site plan and landscaping plans, are numbered separately from the architectural drawings. If this is the case, the architectural drawings are sequentially numbered in order of basement or ground-floor plans; upper-level floor plans; exterior elevations; sections; interior elevations; details; and window, door, and room finish schedules.

2.1.1 Site Plans

The main purpose of the site plan (also called a *civil plan*) is to locate the structure within the confines of the building lot. The site plan clearly shows the building's dimensions (*Figure 12*). The building is usually shown by the size of the foundation and the distances to the respective property lines.

A commercial building needs a site that is appealing, convenient for customer traffic, and in harmony with other commercial structures in the area. Commercial site drawings show many of the same features as residential drawings but are far more detailed. They include details on site improvement features, existing and finish **contour lines**, paving areas, and site access. Because a commercial structure serves the public, the architect, owner, and zoning board should note and specify all site changes. Commercial site plans will also show the location of public utilities.

When looking at a commercial site plan for the first time, notice the features that also appear on residential location plot plans, including the following:

- Survey data such as directional arrow, property lines, and the structure's relationship to other features
- Geographic data such as lot corner elevations, existing and proposed contours, and landscaping features
- Building features such as floor elevations, exterior wall positioning, roof overhang, and drainage

Features on commercial site plans that may not be found on residential plans include the following:

- Details of paved areas, including walkways, driveways, and parking areas
- Details on grading and other site work

Contour lines: Imaginary lines on a site plan/plot plan that connect points of the same elevation. Contour lines never cross each other.

Figure 12 Site plan.

- Notations that refer to details on the site plan or elsewhere in the drawings
- Positions of existing and new utility lines
- Dimensions, usually specified in feet and tenths of feet
- Referenced details, sections, and elevations pertaining to site preparation features
- Symbols or notations detailing material types, sizes, and positions

Another primary purpose of the site plan is to show the unique surface conditions, or topography, of the lot. The topography of a particular lot may be shown right on the site plan. For projects in which the topography must be shown separately for clarity, a grading plan is used.

Did You Know?

Museum of American Architecture

The Athenaeum of Philadelphia is a museum of American architecture and interior design. The work of over 1,000 American architects from 1800 to 1945 is collected and available for research. The museum holds 150,000 drawings, 50,000 photographs, and many other documents. Most of the drawings are from Philadelphia, but the collection also includes drawings of buildings in most of the states and several countries. The building was designed in 1845 by John Notman. It was one of the first Philadelphia buildings built of brownstone.

Source: Justin, CC, via Wikimedia Commons

2.1.2 Floor Plans

Floor plans of both commercial and residential structures show various floor levels as if a horizontal plane had been cut through the structure. This results in an overhead view of each floor. One common use of the floor plan is to calculate the area of each room and other areas represented on the floor plan. This information is needed to determine the amount of floor covering, wallboard, and other material required. For most rooms, it is simply a matter of multiplying the length and width dimensions to determine the square footage of the room. If the room is not a perfect rectangle, however, the calculation must be done by breaking the room into pieces, then adding the square footage of the pieces.

The most noticeable difference between commercial and residential floor plans is the amount of detail. Commercial plans contain more details of room use, finishes, wall types, sound transmission, and fire retardation. In fact, most commercial floor plans will incorporate a legend or chart to specify the various interior wall types shown on the plan. The drawing scale is normally $\frac{1}{8}$ inch to 1 foot for the plan views. The detailing instructions usually include the following types of information:

- Revisions and changes that may be enclosed by a cloud outline
- Numerous **callouts** specifying sectional views and details
- Room assignment designations by function or number
- Detailed dimensioning of all visible parts of the structure
- Finish designations referenced to schedules
- Life safety drawings, which include requirements for such things as sprinkler systems, fire walls, and exit corridors

Callouts: Markings or identifying tags describing parts of a drawing; callouts may refer to detail drawings, schedules, or other drawings.

Color Codes

It is a good idea to color code different wall types on a floor plan using different colored markers. In addition to simplifying the installation, it will also ensure that you review the installation line by line, which provides an opportunity to plan the installation and anticipate any problems.

On commercial plans, little is left to chance for several reasons. First, the construction itself is varied and complex. Second, construction must meet all code specifications. Finally, different contractors will be working on the same job. Because there are many ways to accomplish the same task, the requirements are specified in detail to achieve consistency throughout the structure.

Each door or window on a floor plan for a commercial building is typically accompanied by a number, letter, or both. This number/letter is an identifier that refers to a door or window schedule that describes the corresponding door or window by size, type of materials, or model number for the specific door or window. Schedules are discussed in more detail later in this section.

When supplied, roof plans (*Figure 13*) provide information about the roof slope, roof drain placement, and other pertinent information. Where applicable, the roof plan may also show information on the location of air conditioning units, exhaust fans, and other ventilation equipment.

To help clarify ambiguous parts of the drawing, notes may be written on commercial floor plans. This is particularly true when any feature differs from one area to another. In most instances, these notes are important not only because they show variations, but also because they detail the responsibilities of those involved in the construction. A plan note may read: "Furnished by owner," "Refer to structural drawings," or "Not in contract." *Figure 13* includes a note instructing the contractor to coordinate the placement of gutters and downspouts with the architect. Be sure to read all notes, even if they apply to other trades. In addition, review all plans, including those of the other trades. They may contain information that pertains to your work.

For large buildings, the architect/engineer often divides the floor plan into sections by grid lines. The grid is the same for all floors of the building. Using the grid allows the architect and engineer to locate features very specifically anywhere in the building. The grid lines are useful for locating features that are repeated on one floor or from one floor to another. The grid markings on the floor plan also reappear on the structural drawings, where they locate footings and columns.

2.1.3 Reflected Ceiling Plans

Imagine you have been tasked with installing the drywall or acoustical panels on a ceiling in a new commercial structure. Visualizing the locations of the light fixtures, sprinkler system, air diffusers, return grilles, sound system, and any other components you need to work around can be problematic if these items are shown on the ceiling. Unless you are holding the plans above your head to view them, your brain has to represent everything on the ceiling reversed, as in a mirror image.

To solve this problem, designers often create a plan that shows the components of the ceiling on the floor. Think of it like covering the floor with a mirror. You would see everything on the ceiling as you are looking at the floor. This view of the ceiling is known as a *reflected ceiling plan*, and *Figure 14* is an example. These plans are often included in architectural drawings to make them more user-friendly.

2.1.4 Schedules and Details

Because a commercial building has so many varied parts and functions, it is impossible to draw all the features on one drawing. Many of the smaller detail features are in notes on the actual drawings. Even then, the plans may become cluttered if too many notes appear. To prevent such a problem, schedules, detail drawings, and written specifications are used to describe the construction materials and procedures required.

In most cases, the architectural drawings will include schedules for doors, windows, and interior wall finishes (*Figure 15*). A schedule is not a list of events and what time they will happen as in the typical sense of the word. In commercial drawings a schedule refers to a chart or table which provides detailed information corresponding to various parts of the drawings. Common features include the following:

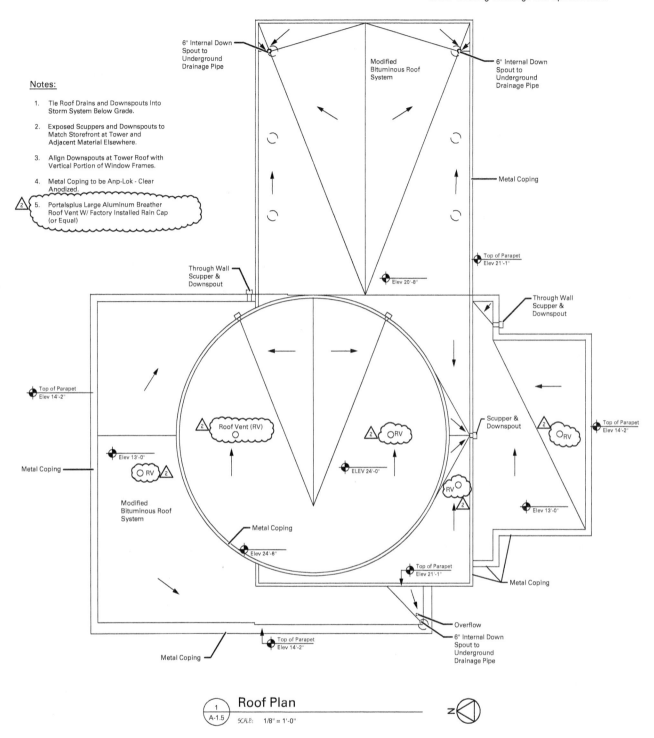

Figure 13 Roof plan.

Source: Photo courtesy of Sanders & Associates Architectural Services

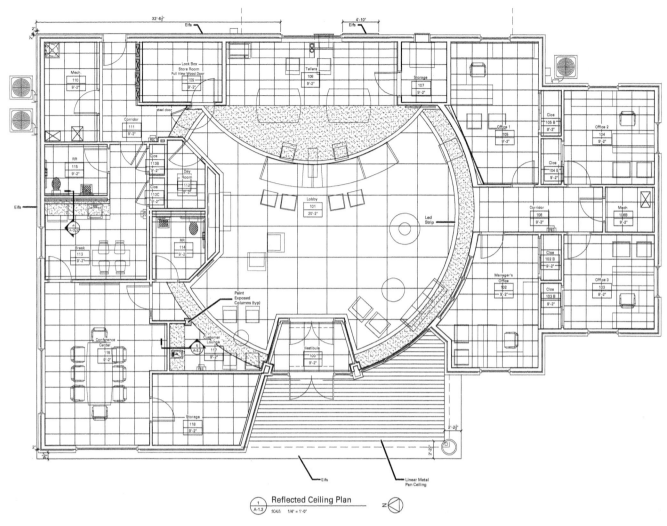

Figure 14 Reflected ceiling plan.

Source: Photo courtesy of Sanders & Associates Architectural Services

DOOR AND FRAME SCHEDULE

	Door						Frame							Fire Rating Label	Notes
	Opening Size									Detail					
Mark	WD	HGT	THK	Matl	Finish	Elev.	Glazing	Matl	Finish	Elev.	Head	Jamb	Sill		
100	6'-0"	7'-0"	1 3/4"	Alum	Anodized	D1	Full	Alum	Anodized	E	1H	1J	--	--	Closer Panic Hdw - Ada Opener
101	6'-0"	7'-0"	1 3/4"	Alum	Anodized	D1	Full	Alum	Anodized	F	2H	2J		--	Closer Panic Hdw - Ada Opener
102A	3'-0"	7'-0"	1 3/4"	Wood	Stained	D2	--	HM	Painted	M	3H	3J		--	--
102B	5'-0"	7'-0"	1 3/4"	Wood	Stained	D5	--	MTL.	-	N	--	--		--	Versatrac Frame (or Equal) Trim as Req'd
103A	3'-0"	7'-0"	1 3/4"	Wood	Stained	D2	--	HM	Painted	M	3H	3J		--	--
103B	5'-0"	7'-0"	1 3/4"	Wood	Stained	D5	--	MTL.	-	N	--	--		--	Versatrac Frame (or Equal) Trim as Req'd
104A	3'-0"	7'-0"	1 3/4"	Wood	Stained	D2	--	HM	Painted	M	3H	3J		--	--
104B	5'-0"	7'-0"	1 3/4"	Wood	Stained	D5	--	MTL.	-	N	--	--		--	Versatrac Frame (or Equal) Trim as Req'd
105A	3'-0"	7'-0"	1 3/4"	Wood	Stained	D2	--	HM	Painted	M	3H	3J		--	--
105B	5'-0"	7'-0"	1 3/4"	Wood	Stained	D5	--	MTL.	-	N	--	--		--	Versatrac Frame (or Equal) Trim as Req'd
106	3'-0"	7'-0"	1 3/4"	Wood	Stained	D2	--	HM	Painted	M	3H	3J		--	--
107	3'-0"	7'-0"	1 3/4"	Wood	Stained	D2	--	HM	Painted	M	4H	4J		1 HR	--
109	3'-0"	7'-0"	1 3/4"	Wood	Stained	D2	--	HM	Painted	M				2 HR	Wood Door with Interior Day Gate - Specialty Hardware Required
110	3'-0"	7'-0"	1 3/4"	Wood	Stained	D2	--	HM	Painted	M	3H	3J		1 HR	--
111	3'-0"	7'-0"	1 3/4"	Steel	Painted	D3	--	2	HM	Painted	Q	3H (SIM)	3J		--	Closer Panic Hdw 1 \ 4
112	3'-0"	7'-0"	1 3/4"	Wood	Stained	D2	--	HM	Painted	M	3H	3J		--	--
113A	3'-0"	7'-0"	1 3/4"	Wood	Stained	D2	--	HM	Painted	M	3H	3J		--	--
113B	3'-0"	7'-0"	1 3/4"	Wood	Stained	D2	--	HM	-	O	--	--		--	--
113C	3'-0"	7'-0"	1 3/4"	Wood	Stained	D2	--	HM	-	O	--	--		--	--
113D	3'-0"	7'-0"	1 3/4"	Wood	Stained	D2	--	HM	Painted	M	3H	3J		--	--
114	3'-0"	7'-0"	1 3/4"	Wood	Stained	D2	--	HM	Painted	M	3H	3J		--	Closer
115	3'-0"	7'-0"	1 3/4"	Wood	Stained	D2	--	HM	Painted	M	3H	3J		--	Closer
116	3'-0"	7'-0"	1 3/4"	Alum	Anodized	D4	Full	Alum	Anodized	K	2H	2J		--	--
118	3'-0"	7'-0"	1 3/4"	Wood	Stained	D2	--	HM	Painted	M	3H	3J		--	--

Figure 15 Door and frame schedule.

Source: Photo courtesy of Sanders & Associates Architectural Services

- A reference mark, letter, or number that corresponds to the markings on the drawings
- The desired manufacturer for a specific item
- Information on item part numbers, sizes, special finishes, and hardware requirements

Door and window schedules are designated by a number or letter on the drawings. The same letter or number is duplicated in the schedule, with a brief description of the item. Typical door and window schedules must show their relationship to the floor plan.

A finish schedule (*Figure 16*) references a location by a room number noted on the schedule; the drawing usually references the schedule with a symbol or note. Finish schedules for commercial buildings provide detailed lists of finishing materials and their application, plus installation procedures for floors, walls, base trim, ceilings, and molding.

ROOM FINISH SCHEDULE

Room No.	Room Name	Floors		Base	Walls								Ceiling	Note
					North		East		South		West			
100	Vestibule	F2	F8	B2	P1		P1		P1		P1		C1	
101	Lobby	F2	F3	B2	P1		P1		P1		P1		C2	
102	Manager's Office	F5		B2	P1	P2	P1	P2	P1	P2	P1	P2	C1	⟨1⟩
102B	Closet	F5		B1	P1		P1		P1		P1		C1	
103	Office 3	F5		B2	P2		P2		P2		P2		C1	
103B	Closet	F5		B1	P2		P2		P2		P2		C1	
104	Office 2	F5		B2	P2		P2		P2		P2		C1	
104B	Closet	F5		B1	P2		P2		P2		P2		C1	
105	Office 1	F5		B2	P1	P2	P1	P2	P1	P2	P1	P2	C1	⟨1⟩
105B	Closet	F5		B1	P1		P1		P1		P1		C1	
106	Corridor	F4		B2	P1		P1		P1		P1		C1	
106B	Mechanical	SC		B1	P1		P1		P1		P1		C1	
107	Storage	F7		B1	P1		P1		P1		P1		C1	
108	Tellers	△2 (F2)		B1	B2	P1		P1		P1		P1		C1	
109	Lock Box Store RM.	F7		B1	P1		P1		P1		P1		C1	
110	Mechanical	SC		B1	P1		P1		P1		P1		C1	
111	Corridor	F7		B2	P1		P1		P1		P1		C1	
112	Dayroom	F5		B2	P1		P1		P1		P1		C1	PL
113	Break Room	F7		B1	P1		P1		P1		P1		C1	
113B	Closet	F7		B1	P1		P1		P1		P1		C1	
113C	Closet	F7		B1	P1		P1		P1		P1		C1	
114	Restroom	△7 (F2)		B3	W1	P1	W1	P1	W1	P1	W1	P1	C1	
115	Restroom	(F2)		B3	W1	P1	W1	P1	W1	P1	W1	P1	C1	
116	Conference Center	△F4		B2	P1		P1		P1		P1		C1	⟨2⟩
117	Customer Lounge	△2 (F2	F3)	B2	P1		P1		P1		P1		C1	⟨2⟩
118	Storage	F7		B1	P1		P1		P1		P1		C1	

Figure 16 Finish schedule.
Source: Photo courtesy of Sanders & Associates Architectural Services

Certain areas of a floor plan, elevation, or other drawing may be enlarged for greater clarity. These enlargements are drawn to a larger scale. They are called details (*Figure 17*). Details can be found either on the sheet where they are first referenced or grouped together on a separate detail sheet included in the set of drawings. These drawings are important sources of information for the contractor and craftworker.

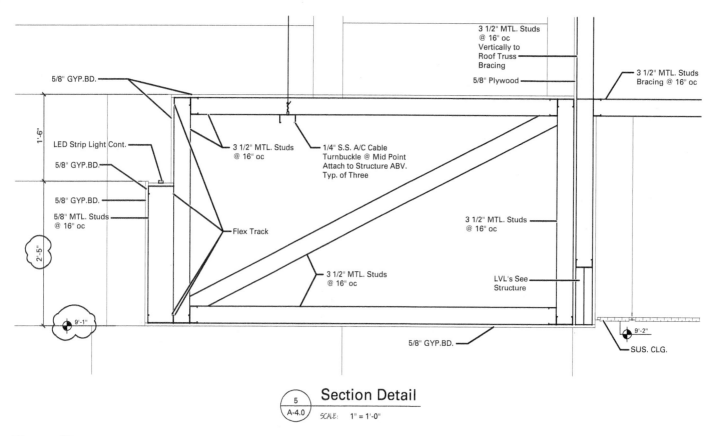

5/8" GYP.BD.

1'-6"

LED Strip Light Cont.

5/8" GYP.BD.

5/8" GYP.BD.

5/8" MTL. Studs
@ 16" oc

2'-5"

9'-1"

3 1/2" MTL. Studs
@ 16" oc

1/4" S.S. A/C Cable
Turnbuckle @ Mid Point
Attach to Structure ABV.
Typ. of Three

Flex Track

3 1/2" MTL. Studs
@ 16" oc

3 1/2" MTL. Studs
@ 16" oc
Vertically to
Roof Truss
Bracing

5/8" Plywood

3 1/2" MTL. Studs
Bracing @ 16" oc

3 1/2" MTL. Studs
@ 16" oc

LVL's See
Structure

9'-2"

SUS. CLG.

5/8" GYP.BD.

5
A-4.0

Section Detail
SCALE: 1" = 1'-0"

Figure 17 Detail drawing.
Source: Photo courtesy of Sanders & Associates Architectural Services

2.1.5 Elevations and Sections

Elevation views on commercial construction drawings are like residential elevations but provide more information. Exterior elevations (*Figure 18*) provide views of the building from each major orientation, as well as references for section views. Elevations are normally drawn to the same scale as the floor plans.

Some interior elevations provide vertical dimensioning for interior work, materials lists, and construction details. They are important for built-in cabinets, shelving, finish carpentry, or millwork items. The scale of the drawing depends on the detail required and may be as small as $\frac{1}{4}$ inch to 1 foot, or as large as $\frac{3}{4}$ inch to 1 foot.

There are many wall sections to show the different types of exterior and interior walls. The interior sections detail the construction of each wall type, such as curtain walls, partitions, load-bearing walls, fire-resistant walls, and noise-reduction walls. The floor plans incorporate a legend that specifies the interior wall type. The wall section drawing provides the necessary detail. Wall sections usually provide the following details:

- Construction techniques and material types
- Stud types and placement
- Fire ratings of various materials measured in terms of hours of resistance
- Sound barrier placement or materials
- Insulation applications and materials
- In-wall features such as recesses or chases

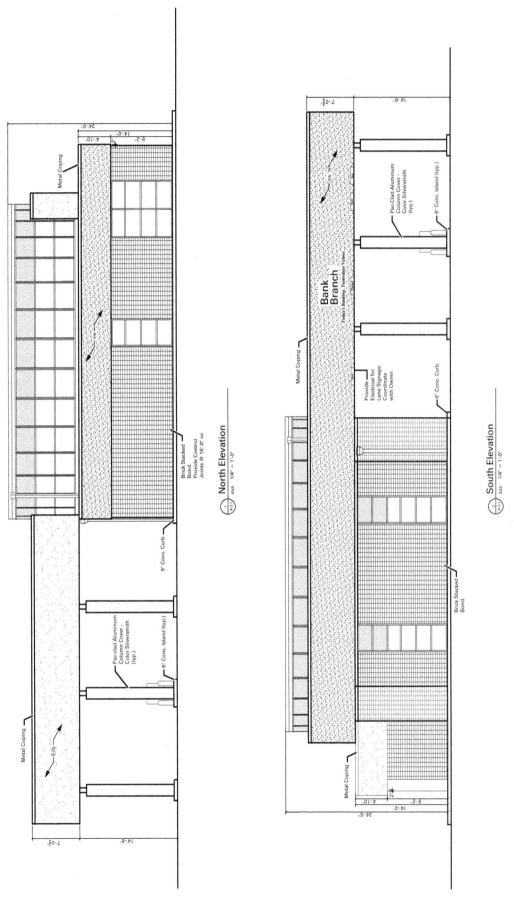

Figure 18 Elevation drawings.

Source: Photo courtesy of Sanders & Associates Architectural Services

Elevations and sections are also drawn for landings, stairways, and back-filled retaining walls. The section drawings detail construction techniques or materials. For instance, a stairway will have details, sections, and elevations with information on tread, landing, and handrail construction. The amount of detail will depend on the complexity of the stairway and on the specifications in relevant building codes.

Longitudinal and transverse wall sections show construction features that are expanded on the structural plans. These sections will show the following:

- Relationships of all wall features from the footings through the roof
- Footing and foundation placement in relation to other elevations
- Exterior materials symbols and notations
- Framework type and placement

As with the elevations, the sections provide an overall view of the proposed structure rather than the detail required for construction. Detail notations or callouts refer directly to the structural plan sheets. Most callouts refer to details on roofing beams and trusses, foundations, framing features, or other structural components.

Building codes: Codes published by state and local governments to establish minimum standards for various types of interior and exterior construction.

Beams: Load-bearing horizontal framing elements supported by walls or columns and girders.

Compressive strength: Refers to a material's ability to support weight.

2.2.0   Structural Drawings

Structural drawings (*Figure 19*) provide detailed information on the structural features of the building. This includes information on the load-bearing design and materials, such as masonry, reinforced concrete, or steel framing. For example, if the floor, roof beams, or trusses place their weight directly on the wall materials, the structure has load-bearing walls. This means that the exterior walls are constructed of materials with high compressive strength such as brick, block, or cast concrete. The walls support their own weight as well as that of the various floor and roof elements. The plans for such load-bearing walls will show no separate supporting structural beams or columns along the wall except for thickened wall sections that support major floor or roof beams. This type of construction is typical in smaller commercial buildings with spans of about 40 feet and heights of no more than three stories.

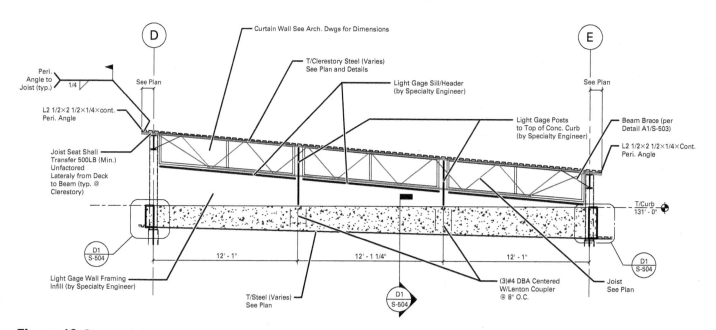

Figure 19 Structural drawing.

In reinforced concrete construction, the load-bearing elements usually include reinforced concrete footings, foundations, piers, beams, floors, columns, and pillars. Various sizes and quantities of carbon steel reinforcement bar (rebar) and/or steel tensioning cable are assembled inside the various building elements before concrete is placed to form the building element. Building exterior cladding is typically nonbearing curtain walls of masonry or other materials.

High-rise buildings are typically steel frame construction with nonbearing exterior curtain walls. The framework for the entire structure is formed by bolting or welding various steel elements together. The loads from floors and roof are transferred to beams and **girders** and down steel columns to the footings. When the joints are bolted, the project schedules designate the number, size, and material requirements for the bolts. *Figure 20* shows the most common shapes for steel frame elements and lists the typical plan designations.

Girders: Large steel or wooden beams supporting a building, usually around the perimeter.

Descriptive Name	Shape	Identifying Symbol	Typical Designation height/wt/ft in lb	Nominal Size height width
Wide Flange Shapes	I	W	W21 × 132	21 × 13
Miscellaneous Shapes	I	M	M8 × 65	8 × 2¼
American Standard Beams	I	S	S8 × 23	8 × 4
American Standard Channels	[	C	C6 × 13	6 × 2
Miscellaneous Channels	[	MC	MC8 × 20	8 × 3
Angles Equal Legs	L	L	L 6 × 6 × ½	6 × 6
Angles Unequal Legs	L	L	L 8 × 6 × ½	8 × 6
Structural Tees (cut from wide flange)	T	WT	WT12 × 73	9
Structural Tees (cut from Am. std. beams)	T	ST	ST9 × 35	9

Figure 20 Structural steel notations.

The structural drawings include plan views, sections, details, schedules, and notes. They provide information on the size and placement of load-bearing elements. They also show how the load-bearing elements are connected to each other and to other parts of the structure.

Typical structural drawings include a foundation plan, floor framing plans, and a roof framing plan (*Figure 21*). The plan view will have sections, details, schedules, and notes located in any available space on the drawing sheet. Each plan view should have a north directional arrow to maintain a consistent orientation. Plan views are typically drawn to the scale of $\frac{1}{8}$ or $\frac{1}{4}$ inch to 1 foot; sections are $\frac{1}{2}$ or $\frac{3}{4}$ inch to 1 foot; details are 1 or $1\frac{1}{2}$ inch to 1 foot.

Each plan is referenced to a grid or checkerboard identifying the placement of columns and/or footings. The grid is the same on all plans. As noted in the section on floor plans, the grid also shows dimensioning lines. In the structural drawings, a callout sequence number or mark will identify and show the placement of columns on the grid. Some project plans do not use a grid but use a callout sequence marking. In either case, the identification marks will be referenced to schedules or notes.

Check the Legend

To avoid mistakes in reading the drawings, be sure you understand the symbols and abbreviations used on every drawing set. Symbols and abbreviations may vary widely from one drawing set to another.

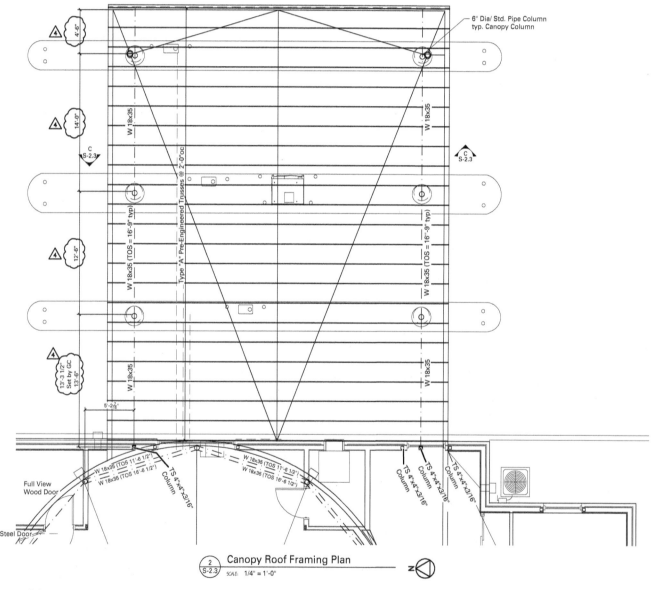

Figure 21 Roof framing plan.

Source: Photo courtesy of Sanders & Associates Architectural Services

2.2.1 Structural Framing Plans

The structural engineer draws a framing plan or diagram for the roof and for each floor level that will be framed. On these drawings, the exterior walls or bearing walls are often lightly drawn in while heavier lines represent the framing. The resulting plan resembles a detailed graph or diagram, as shown in *Figure 22*.

Notice the following when looking at this example, or at any framing plan:

- Notations identifying beams, joists, and girders by size, shape, and material
- Column, pier, and support locations and their relationship to joists or framing
- Notes or callouts identifying corresponding sections or detail drawings

You may also see details for locations of stairs, recesses, and chimneys. Details show additional unique framing around these areas. The dimensions on the drawings are centerline dimensions, not actual member dimensions. The member size must be less than the centerline dimensions to allow for construction tolerances.

The columns on a framing plan are shown from the top. Lines running between columns are beams. Beams fasten directly to the columns. Joists fasten between beams or between beams and walls. To keep long spans from swaying or twisting in the center, bridging or support members are placed between joists. All of these members will either be referenced to a schedule or to notes directly written on the plans.

Joists: Horizontal members of wood or steel supported by beams and holding up the planks of floors or the laths of ceilings. Joists are laid edgewise to form the floor support.

Visualizing the Future with BIM

Since the 1980s, computer-aided drafting (CAD) has been the predominant method for creating and revising construction drawings. An advancement known as Building Information Modeling (BIM), a collaborative 3D modeling and project management tool, is making a big impact on the commercial construction industry. Although BIM is not a replacement for the design capabilities of CAD, it has some powerful advantages for practical jobsite applications. BIM software works by creating three-dimensional representations of the components of a structure and combining them into one holistic 3D model of the entire project. Everyone working on the project can use the same model to visualize their work and the work of others, creating an effective means of collaboration throughout the life of the project. If a change is made to one component, it is updated across the entire model. Some BIM platforms incorporate project management features like cost estimating, scheduling, and quantity takeoffs.

Source: KRAUCHANKA HENADZ/Shutterstock

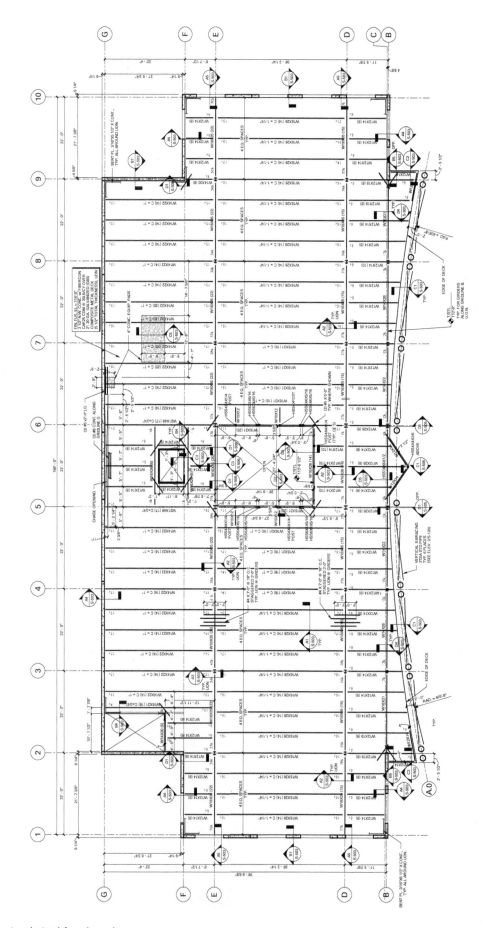

Figure 22 Structural steel framing plan.

Structural drawings include sheets of details, schedules, and notes with the framing plans. They help workers to understand and follow the specifications for the structure, and provide information on the following areas:

- Reinforcing information for all areas where steel rods or wire mesh will be used
- Information for each type of connection made in framing members
- Bearing plate information detailing the features of all members that will bear directly on other members
- Information for positioning ties, stirrups, or saddles
- Placing and construction information for any unique features that cannot be adequately described with a drawing

The details also identify load limits, test strength, fastener types, and uniform specifications, which must be applied where specific information is not given.

Framing plans contain a great deal of information. It is easier to read such plans by isolating the separate bays or spans between columns. Read the details for that area before moving to other areas of the plan.

2.2.2 Concrete Structures

Concrete is one of the most widely used materials in construction today. Concrete's strength and durability make it a desirable construction material. Concrete is typically used in building foundations because it has a high compressive strength. Concrete on its own has low **tensile strength**, which means that concrete can be easily broken apart when it is flexed. When reinforced with steel reinforcing bar, often called rebar, concrete has good tensile strength.

Tensile strength: Refers to the flexibility of a material.

Concrete foundations typically have footings around the edges and wherever support columns are needed. Footings support the weight of the building walls and roof, so they need to be stronger than the building floor. Footings have more steel reinforcement than foundations and are dug much deeper into the ground, usually below the regional frost line. *Figure 23A* shows a footing detail drawing for a concrete foundation. *Figure 23B* is an example of a schedule which gives more information about the footings.

Frequently, the framework for a large construction project will need to use special, heavy-duty reinforcing techniques to strengthen the concrete. When a load is applied to a conventional concrete slab, such as an elevated parking garage deck, the concrete tends to sag and may develop cracks. The use of rebar in the slab combats this problem but is not enough to prevent cracking under heavy loads. To help correct this, post-tensioned and pre-tensioned reinforcement is used.

In post-tensioning, a steel tendon is placed in the form with the ends protruding. The tendon is placed in accordance with a post-tensioning profile drawing developed by a qualified engineer. *Figure 24* is an example of post-tensioning profiles for post-tensioned beams. The multi-span beam profile diagram shows how the tendons go across the top of a column. A tendon consists of a threaded bar or strands of wire, along with anchoring hardware and sheathing. The strand is typically made from seven steel wires, each $1/2$ inch in diameter, which have been twisted together. Once the concrete has hardened around the tendon, one end is anchored, and the other end is tensioned using a special hydraulic jack. When the desired tension has been reached, the tendon is secured.

The two types of post-tensioning are bonded and unbonded. In bonded post-tensioning, a steel or plastic duct is inserted in the form. After the concrete has been poured and has hardened, the strand is threaded through the duct. Tensioning is applied, and then the duct is filled with grout.

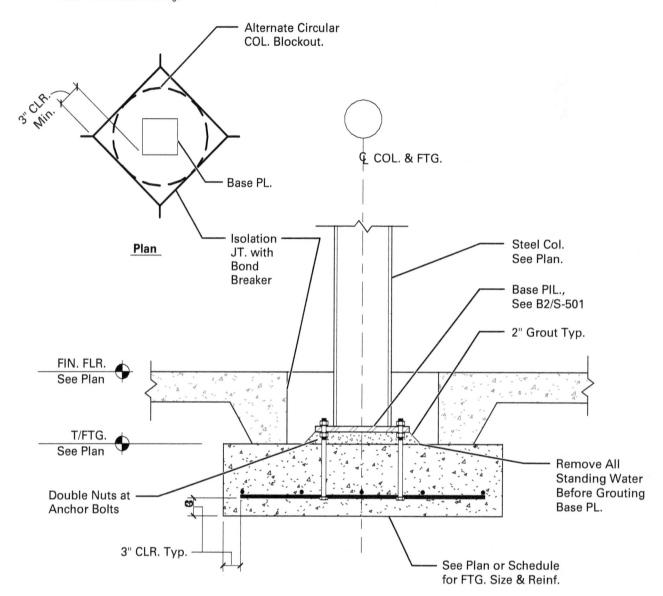

Alternate Circular
COL. Blockout.

3" CLR. Min.

Base PL.

Plan

Isolation
JT. with
Bond
Breaker

℄ COL. & FTG.

Steel Col.
See Plan.

Base PIL.,
See B2/S-501

2" Grout Typ.

FIN. FLR.
See Plan

T/FTG.
See Plan

Double Nuts at
Anchor Bolts

3" CLR. Typ.

Remove All
Standing Water
Before Grouting
Base PL.

See Plan or Schedule
for FTG. Size & Reinf.

Figure 23A Footing detail drawing.

FOOTING SCHEDULE

Mark	Width	Length	Thickness	Reinforcing	Remarks
CF2.0	2' - 0"	CONT.	1' - 0"	2-#5 CONT. w/ #3 @ 48" TRANS.	
CF2.5	2' - 6"	CONT.	1' - 0"	2-#5 CONT. w/ #3 @ 48" TRANS.	
CF3.0	3' - 0"	CONT.	1' - 0"	2-#5 CONT. w/ #3 @ 48" TRANS.	
CF3.5	3' - 6"	CONT.	1' - 0"	2-#5 CONT. w/ #3 @ 48" TRANS.	
F6.0	6' - 0"	6' - 0"	1' - 2"	6-#6 EA. WAY BOT.	Rectangular Footings
F6.0T	6' - 0"	6' - 0"	1' - 2"	6-#6 EA. WAY TOP & BOT.	Rectangular Footings
F7.0	7' - 0"	7' - 0"	1' - 6"	7-#6 EA. WAY BOT.	Rectangular Footings
F7.5	7' - 6"	7' - 6"	1' - 6"	8-#6 EA. WAY BOT.	Rectangular Footings
RF6x8	6' - 0"	8' - 0"	1' - 2"	#6 EA. @12"o/c EA. WAY TOP & BOT.	Rectangular Footings
RF6x9	6' - 0"	9' - 0"	1' - 2"	#6 EA. @12"o/c EA. WAY TOP & BOT.	Rectangular Footings
RF6x11.5	6' - 0"	11' - 6"	1' - 2"	#6 EA. @12"o/c EA. WAY TOP & BOT.	Rectangular Footings

Figure 23B Footing schedule.

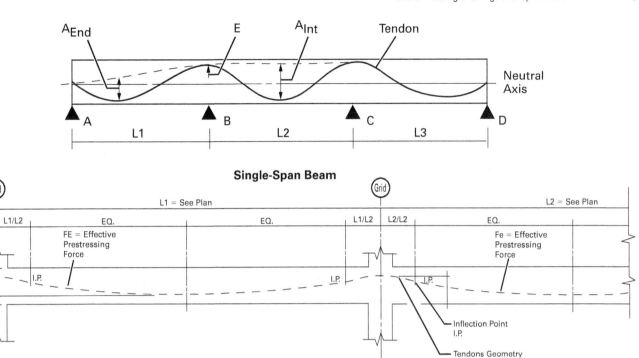

Figure 24 Examples of tendon profiles.

Figure 25 shows post-tensioning tendons protruding from recently poured concrete. Once the tendons have been anchored, they are cut, and the openings are sealed (*Figure 26*).

Figure 25 Post-tensioning tendons.
Source: DCA88/Shutterstock

Figure 26 Sealed tendon openings.
Source: Industria/Alamy Images

In an unbonded system, the strand is covered with corrosion-inhibiting grease and encased in a waterproof plastic sheath. The entire assembly is placed into the form before the concrete is poured.

The twisted-wire strand is used in large structures. The threaded bar is more common in smaller structures; a bearing plate and nut are used to anchor the bar.

WARNING!

Do not drill or install powder-actuated fasteners into any type of post-tensioned concrete because you might damage the tendon, which may fly out of the slab, causing injury. Damaged tendons also reduce the load-bearing capacity of the beam or slab.

Pre-tensioning uses the same techniques as post-tensioning except that the components of the structure are manufactured and tensioned at the manufacturing site. Then, the entire assembly is transported to the construction site for installation.

The interior walls of concrete construction are typically framed with steel studs, but furring strips or full-size wood studs may also be used. These allow a means to install thermal and moisture barriers, electrical wiring, and drywall panels.

2.3.0 Mechanical, Electrical, and Plumbing (MEP) Drawings

Mechanical, electrical, and plumbing plans are referred to collectively as MEP plans. Depending on the size and complexity of a commercial construction project, mechanical, electrical, and plumbing systems will have separate plan sets.

As a drywall technician, you will not work directly on these systems, but in many situations you will be required to make passage for or work around MEP items. One example would be the through-wall ductwork for a forced-air HVACR system. Because of these types of requirements, you must carefully review all MEP drawings for information on dimensions and measurements. Locate services entering the building and passing through interior walls and coordinate with any HVACR, electrical, or plumbing professionals working on site.

2.3.1 Mechanical Drawings

Mechanical drawings provide information about the heating, ventilation, air conditioning, and refrigeration systems (HVACR). The work required to install these systems in commercial structures is typically performed by specialized craft workers.

Mechanical plans typically contain schedules that identify the different types of HVACR equipment. As appropriate, the plans will include a detailed view describing the installation of the HVACR equipment. Depending on the nature of the project, these views can include a refrigeration piping schematic, chilled-water coil and hot-water coil piping schematics, and piping runs for other HVACR equipment.

A basic understanding of the air handling system is a primary requirement for reading an HVACR drawing. In a forced-air system, fans move the heated or cooled air through ductwork into the working areas, offices, and public spaces. Air return ducts collect air from these areas and channel it back through the system or to the outside atmosphere.

Figure 27A, *Figure 27B* and *Figure 27C* include three drawings for HVACR systems. The first shows a typical mechanical drawing for an HVACR system, which involves laying out all the piping, controllers, and air handlers to scale on a basic floor plan. Most of the details normally on a floor plan have been removed so the details of the HVACR system can be easily seen.

The second drawing, shown in *Figure 27B*, includes the following:

- Layout of all the water supply and return and air handling units
- Piping diagrams for the hot-water and chilled-water coils
- Legend that identifies the abbreviations used in the drawings

The third drawing, *Figure 27C*, is a section of a typical shop drawing showing ductwork details for a project. The ductwork for these systems is usually one of the following types:

- Individual duct system
- Trunk duct system
- Crawl space plenum or plenum ceiling systems

The first two delivery systems may be used either from the floor or ceiling levels. The ductwork begins at the central unit and moves out to each area of the workspace. The plans may specify circular or rectangular ducts. Circular ducts slip together and bend to form angles. Rectangular ducts fit between joists and studs. In the individual duct system, many small ducts may reach from the central unit to each room, where a grilled register delivers the air. In a trunk duct system, large main ducts run the length of the structure and smaller ducts branch off to each room register.

In some forced-air systems, particularly in buildings with a crawl space, the heating unit is designed to heat the exterior walls and the floor above the crawl space. A crawl space plenum system heats the entire crawl space, and the conditioned air rises through floor registers. This system requires no ductwork to transfer the warm air because it is transferred through the structural frame.

To gain the most information from HVACR plans, you should look for the following:

- The direction of the central fan
- Duct supply lines, which deliver the conditioned air to all parts of the structure
- Return duct (some HVACR systems have return ducts to return air to the central unit for more heating or cooling)

Mechanical drawings that contain HVACR systems should be carefully studied for the layout, locations, and sizes of ductwork. Special mechanical drawings that illustrate the components of boiler and rooftop- or grade-mounted HVACR units may also be included.

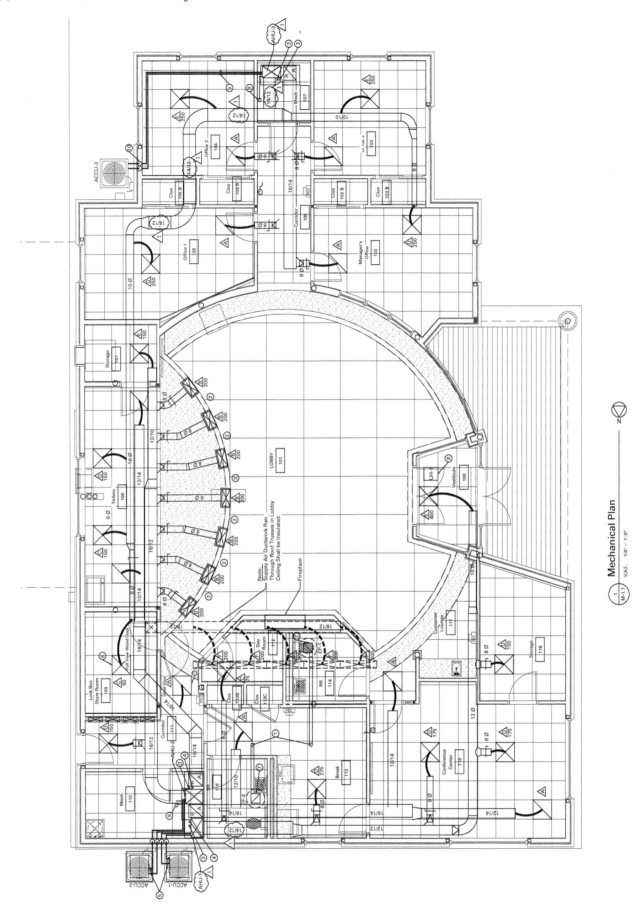

Figure 27A HVACR mechanical plan (1 of 3).

Source: Photo courtesy of Sanders & Associates Architectural Services

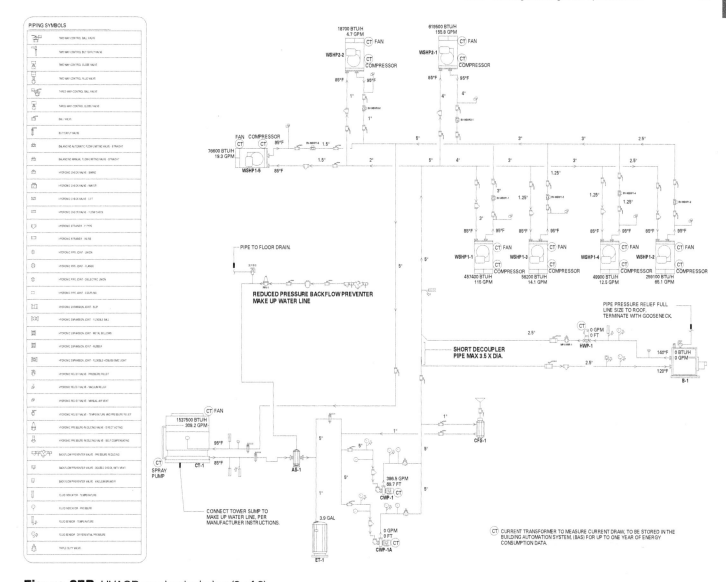

Figure 27B HVACR mechanical plan (2 of 3).

Coordination Drawings

Coordination drawings are produced by the individual contractors for each trade to prevent a conflict in the installation of their materials and equipment. Coordination drawings are produced prior to finalizing shop drawings, cut lists, and other drawings, and before the installation begins. The development of these drawings evolves through a series of review and coordination meetings held by the various contractors.

Some contracts require coordination drawings, while others only recommend them. When one contractor elects to make coordination drawings and another does not, the contractor who made the drawings may be given the installation right-of-way by the presiding authority. As a result, the other contractor may have to bear the expense of removing and reinstalling equipment if it was installed in a space designated for use by the contractor who produced the coordination drawings.

2.3.2 Electrical Drawings

Electrical drawings identify the layout of the electrical distribution system, the lighting requirements, and the telecommunications and computer connections. This information is provided in plain view on a copy of the floor plan.

Although you do not need to understand the details of electrical circuitry to read an electrical plan, you should understand the basic design of the system. All systems begin at the service source. Like water and natural gas, the process

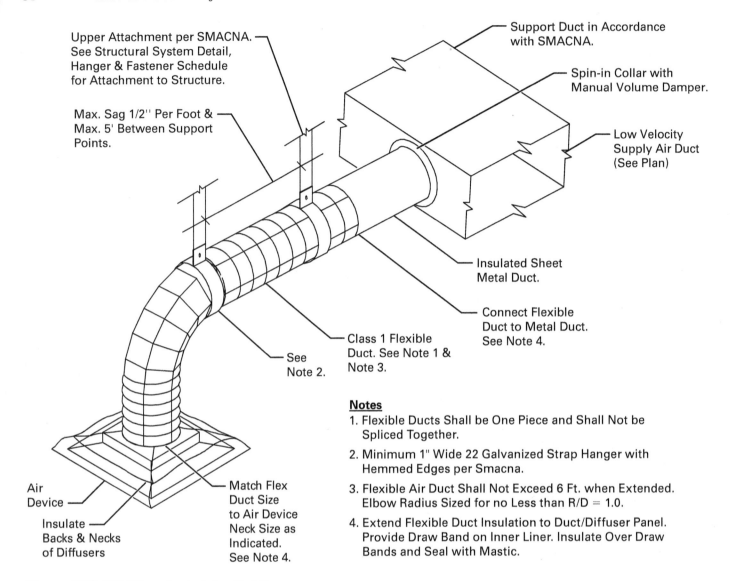

Upper Attachment per SMACNA. See Structural System Detail, Hanger & Fastener Schedule for Attachment to Structure.

Max. Sag 1/2" Per Foot & Max. 5' Between Support Points.

Support Duct in Accordance with SMACNA.

Spin-in Collar with Manual Volume Damper.

Low Velocity Supply Air Duct (See Plan)

Insulated Sheet Metal Duct.

Connect Flexible Duct to Metal Duct. See Note 4.

Class 1 Flexible Duct. See Note 1 & Note 3.

See Note 2.

Air Device

Insulate Backs & Necks of Diffusers

Match Flex Duct Size to Air Device Neck Size as Indicated. See Note 4.

Notes
1. Flexible Ducts Shall be One Piece and Shall Not be Spliced Together.
2. Minimum 1" Wide 22 Galvanized Strap Hanger with Hemmed Edges per Smacna.
3. Flexible Air Duct Shall Not Exceed 6 Ft. when Extended. Elbow Radius Sized for no Less than R/D = 1.0.
4. Extend Flexible Duct Insulation to Duct/Diffuser Panel. Provide Draw Band on Inner Liner. Insulate Over Draw Bands and Seal with Mastic.

Figure 27C HVACR mechanical plan (3 of 3).

begins with a meter being installed by the utility company. From the meter, service moves through a main cutoff switch to the service entry panel. The panel separates service into branch circuits. These circuits are protected from overload by circuit breakers with various capacities. The branches then feed out to specific fixtures and outlets throughout the building.

The complexity of electrical systems, like plumbing systems, varies depending on the structure being constructed. The electrical drawings for small warehouses and office buildings will be simple in comparison to those of a research laboratory or hospital. Some complex facilities have backup generators, parallel wiring systems, alarm and security systems, audio visual systems, and other components of which drywall technicians and steel framers need to be aware.

Power Plan

A typical power plan illustrates the power requirements of the structure. *Figure 28* is an example of a power plan showing the locations of outlets and circuitry. *Figure 29* is a part of the power plan called a riser diagram and would normally show the panels, receptacles, and circuitry of power-utilizing equipment. Some buildings need multiple panels. The power plan would then include panel schedules. Panel schedules list the circuits in the panel, the individual power required for each panel, and the total power requirement for the system. The schedules

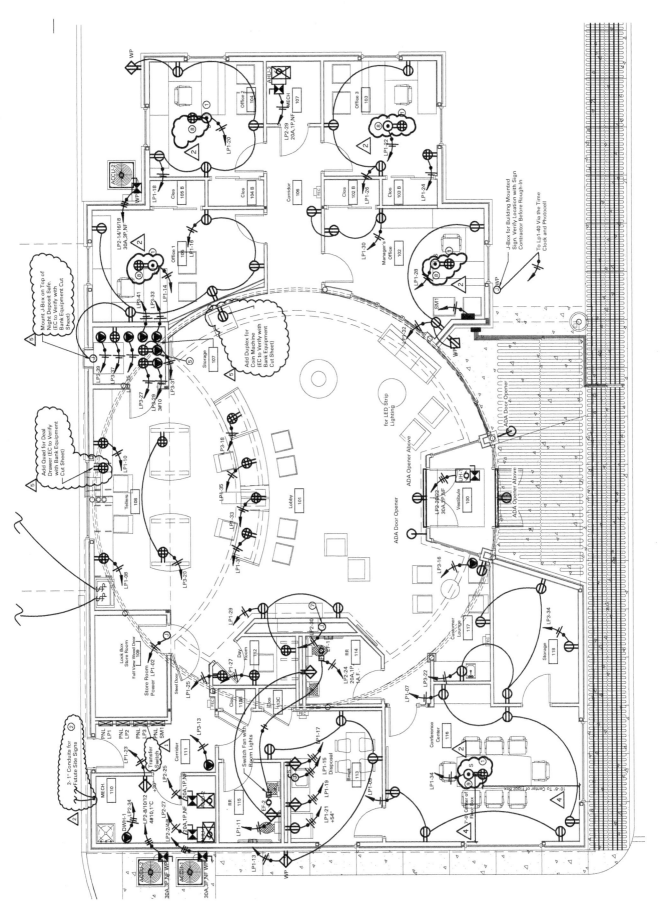

Figure 28 Power plan.

Source: Photo courtesy of Sanders & Associates Architectural Services

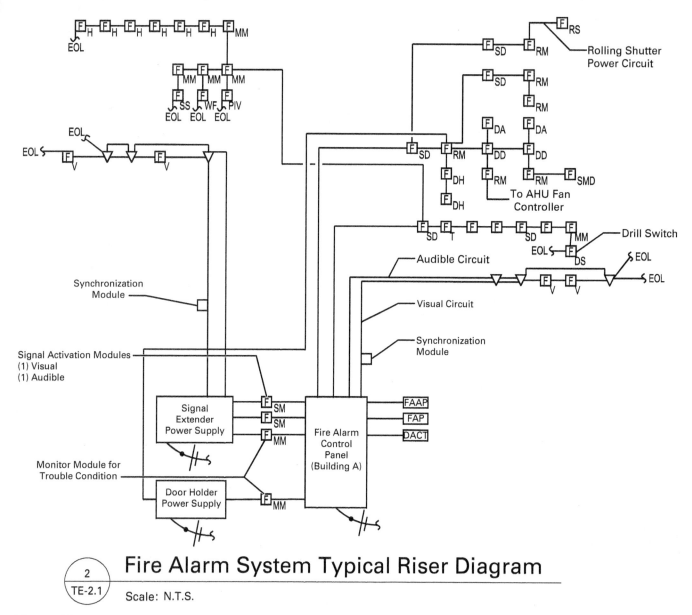

Fire Alarm System Typical Riser Diagram

2
TE-2.1   Scale: N.T.S.

Figure 29 Riser diagram for fire alarm system.
Source: Photo courtesy of Sanders & Associates Architectural Services

total the power requirements to help the electrical contractor size the panel. The power company sizes the overall service requirements.

Lighting Plan

A typical lighting plan locates the various lighting fixtures in the building (*Figure 30*). It is complemented by a fixture schedule that lists the types of light fixtures to be used. The fixture schedule is organized by number or letter. These numbers or letters refer to the manufacturer and model, the wattage of the lamps, voltage, and any special remarks concerning the fixture. Outlets vary according to their electrical capacity and type. Different symbols designate outlets with different voltage requirements. This is also true of most other types of electrical items. Details such as type, size, and voltage are listed on the schedules.

The lighting plan can also include smoke- and fire-detection equipment, emergency lighting, cable TV, and telephone outlets. Buildings with more complex electrical systems may include separate plans for fire prevention and tele-communications systems.

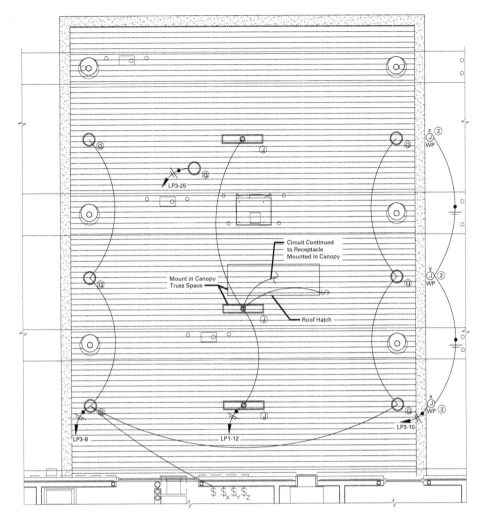

Figure 30 Lighting plan.
Source: Photo courtesy of Sanders & Associates Architectural Services

Did You Know?

Lighting the New York Skyline

The first searchlight on top of the Empire State Building heralded the election of Franklin D. Roosevelt in 1932. A series of floodlights were installed in 1964 to illuminate the top 30 floors of the building. Today, the color of the lights is changed to mark various events. Yellow marks the US Open Tennis Championships, and red, white, and blue mark Independence Day.

The building is lit from the 72nd floor to the base of the TV antenna by 204 metal halide lamps and 310 fluorescent lamps. In 1984, a color-changing apparatus was added in the uppermost mooring mast. There are 880 vertical and 220 horizontal fluorescent lights. The colors can be changed with the flick of a switch.

On Earth Day and Rainforest Awareness Day, the Empire State Building's tower lights are green.

2.3.3 Plumbing Drawings

All buildings that will be occupied require some type of plumbing. Warehouses have simple toilets, while hospitals and restaurants have sophisticated plumbing systems. As with fire protection, plumbing work requires special knowledge and training. Plumbing work is usually performed by a licensed craftsperson. A separate permit and inspection process is required.

A plumbing drawing, as shown in *Figures 31A* and *Figure 31B*, usually appears as a plan view drawing and as an **isometric drawing** called a riser diagram. The

Isometric drawing: A three-dimensional type of drawing in which the object is tilted so that all three faces are equally inclined to the picture plane.

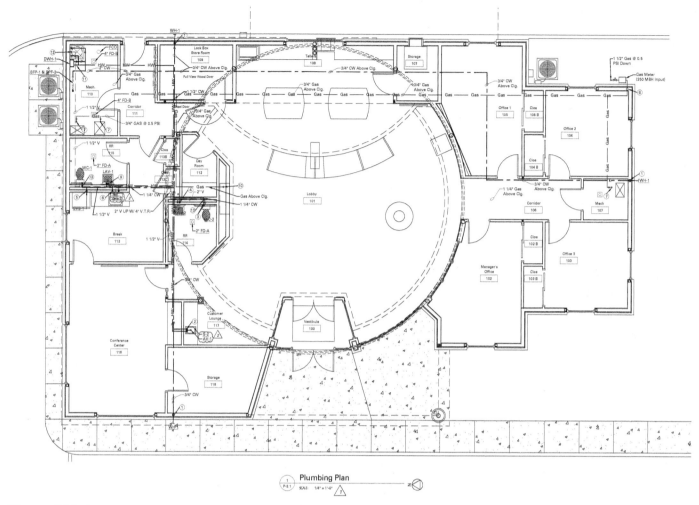

Plumbing Plan

Figure 31A Plumbing plan view.

Source: Photo courtesy of Sanders & Associates Architectural Services

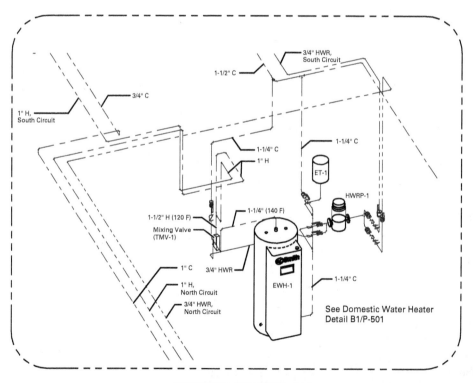

Riser Diagram - Water Heating System

Figure 31B Plumbing riser diagram.

plan view shows the horizontal distances or piping runs. The riser diagram shows vertical pipes in the walls. Plumbing drawings for water systems usually show two separate systems: a water source or distribution system and a waste collection and disposal system.

The incoming water systems operate under pressure. The distribution system shows the hot- and cold-water supply lines. Distribution plans usually include several of the following types of information:

- Piping type and size: types include copper, plastic, brass, iron, or steel piping; the sizes usually appear as nominal dimensions approximating the inside diameter (ID) of the pipe
- Pipe fitting at joints, which includes couplings (straight-run joints), elbows (45- or 90-degree bends), tees, and valves (also called cocks, bibbs, or faucets)
- Cold water supply piping from the street main to the service meter and on to the various fixtures
- Hot water supply piping locations from the hot water heater to the various fixtures; in many instances, piping includes a hot water return line
- A legend identifying cold- and hot-water supply lines

The waste system works using gravity, and it must be vented. The waste system is known as drain, waste, and vent, or DWV piping. When reading waste disposal plans, you can expect to see a combination of several notations. You will usually find the following information on a plumbing drawing for a waste disposal system:

- Waste disposal line size
- Location of fixture branches (horizontal disposal pipes)
- Stack locations (vertical disposal pipes are called waste stacks if they carry toilet waste; soil stacks if they carry waste materials other than toilet waste; and vent stacks if they extend through the roof to release gases within the stacks)
- Locations for drains, sewer lines, cleanouts, traps, meters, and valves

Because waste disposal occurs using the force of gravity, fixture branches are sloped between $\frac{1}{3}$ and $\frac{1}{2}$ inch per foot away from the fixtures.

Piping for distribution of natural gas within the structure is often considered part of the plumbing work. The utility company installs a meter where natural gas enters the building. The piping starts at the meter. Natural gas is distributed to the gas-fueled appliances within the building such as boilers, furnaces, water heaters, and rooftop HVACR units.

Typically, black steel pipe with threaded fittings is used to distribute natural gas. The fittings are like those of other piping systems. They include elbows, bends, unions, and tees. Valves to control the flow of gas within the pipe are typically brass. Gas systems may require openings through masonry walls to accommodate piping, valves, and regulators. Through-wall piping must not be rigidly connected to the wall due to the different rates of expansion and contraction between the piping and the masonry.

Sprinkler Systems

Sprinkler contractors use the architectural and structural drawings of the project to clarify the requirements for any sprinkler system installation. The basic sprinkler system drawings for a building are plan view drawings that contain most of the information for a sprinkler system layout. Reflected ceiling plans typically include sprinkler head locations. From this information, drawings are developed to show the detailed data necessary for an installation. *Figure 32* is an example of a sprinkler system drawing showing the detailed sprinkler system layout for the first and second floors of an office building.

Did You Know?

The Empire State Building

The Empire State Building has 102 stories. It has 70 miles of water pipe that provide water to tanks at various levels. The highest tank is on the 101st floor. There are two public restrooms on each floor and a number of private bathrooms. There are, however, no water fountains.

Isometric Drawings

Isometric means equal measurement. A designer uses the true dimension of an object to construct the drawing. An isometric drawing shows a three-dimensional view of where the pipes should be installed. The piping may be drawn to scale or to dimension, or both. In an isometric drawing, vertical pipes are drawn vertically on the sketch, and horizontal pipes are drawn at an angle to the vertical lines.

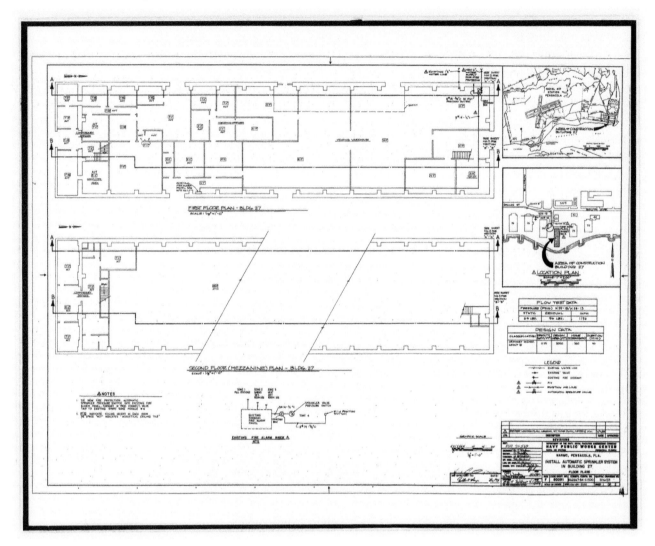

Figure 32 Sprinkler system drawing.
Source: Tango Images/Alamy Images

2.4.0 Specifications

The specifications for a building or project are the descriptions in writing of work and duties required of the owner, architect, and consulting engineer. Together with the working drawings, these specifications form the basis of the contract requirements for the construction of the building or project. Those who use the construction drawings and specifications must always be alert to discrepancies between the working drawings and the written specifications. The following are some situations where discrepancies may occur:

- Architects or engineers use standard or prototype specifications and attempt to apply them without any modification to specific working drawings.
- Previously prepared standard drawings are changed or amended by reference in the specifications only, and the drawings themselves are not changed.
- Items are duplicated in both the drawings and specifications, but an item is subsequently amended in one and overlooked in the other contract document.

Legally, specifications take precedence over drawings. However, the person in charge of the project has the responsibility to determine whether the drawings or the specifications take precedence in a given situation.

2.4.1 How Specifications Work

Writing accurate and complete specifications for building construction is a serious responsibility for those who design the buildings. Specifications, along with the working drawings, govern the entire construction process. They have to be detailed and accurate. Compiling and writing these specifications is not a simple task, even for those who have considerable experience in preparing such documents.

A set of written specifications for a single project usually contains thousands of products, parts, and components, and the methods of installing them, all of which must be covered in either the drawings and/or specifications. No one can memorize all the necessary items required to describe accurately the various areas of construction on a project. The person writing the specifications must rely upon reference materials such as manufacturer's data, catalogs, and checklists to produce high-quality specifications.

Figure 33 is a small sample of a larger ten-page document containing the specifications for drywall for a commercial project. The specifications for the entire document consist of more than 1,200 pages. In order for drywall technicians and other skilled craft workers to locate the specifications that pertain to their work, a universal system of organization is necessary.

2.7   JOINT TREATMENT MATERIALS

A.   General: Comply with ASTM C 475/C 475M.

B.   Joint Tape:

    1.   Interior Gypsum Board: Paper.
    2.   Exterior Gypsum Soffit Board: Paper.
    3.   Glass-Mat Gypsum Sheathing Board: 10-by-10 glass mesh.
    4.   Tile Backing Panels: As recommended by panel manufacturer.

C.   Joint Compound for Interior Gypsum Board: For each coat use formulation that is compatible with other compounds applied on previous or for successive coats.

    1.   Embedding and First Coat: For embedding tape and first coat on joints, fasteners, and trim flanges, use drying-type, all-purpose compound, unless setting-type taping or sandable topping is required or recommended by manufacturer for intended application.

        a.   Use setting-type compound for installing paper-faced metal trim accessories.

    2.   Fill Coat: For second coat, use drying-type, all-purpose compound, unless setting-type taping or sandable topping is required or recommended by manufacturer for intended application.
    3.   Finish Coat: For third coat, use drying-type, all-purpose compound, unless setting-type taping or sandable topping is required or recommended by manufacturer for intended application.

D.   Joint Compound for Tile Backing Panels:

    1.   Glass-Mat, Water-Resistant Backing Panel: As recommended by backing panel manufacturer.
    2.   Cementitious Backer Units: As recommended by backer unit manufacturer.
    3.   Water-Resistant Gypsum Backing Board: Use setting-type taping compound and setting-type, sandable topping compound.

Figure 33 Sample of specifications for drywall.

2.4.2 Format of Specifications

The most suitable organization of the specifications is a series of sections dealing with the construction requirements, products, and activities. Specifications should be easily understandable by the different trades. Those who use them must be able to find all the information needed without spending too much time looking for it.

The most used specification format in North America is *MasterFormat*™. The Construction Specifications Institute (CSI) and Construction Specifications Canada (CSC) jointly developed this standard. For many years prior to 2004, the organization of construction specifications and supplier's catalogs was based on a standard with 16 sections, known as divisions. The divisions and their subsections were individually identified by a five-digit numbering system. The first two digits represented the division number and the next three individual numbers indicated successively lower levels of classification. For example, the number 13213 represented division 13, subsection 2, sub-subsection 1 and sub-sub-subsection 3.

Specifications

Written specifications supplement the related working drawings and contain details not shown on the drawings. Specifications define and clarify the scope of the job. They describe the specific types and characteristics of the components that are to be used on the job and the methods for installing some of them. Many components are identified specifically by the manufacturer's model and part numbers. This type of information is used to purchase the various items of hardware needed to accomplish the installation in accordance with the contractual requirements.

In 2004, the *MasterFormat*™ (*Figure 34*) underwent a major change. What had been 16 divisions was expanded to four major groupings and 49 divisions with some divisions reserved for future expansion. The first 14 divisions are essentially the same as the old format. Subjects under the old Division 15—*Mechanical* have been relocated to new divisions 22 and 23. The basic subjects under old Division 16—*Electrical* have been relocated to new divisions 26 and 27. As a drywall technician, Division 9—*Finishes* will be your primary division.

In addition, the numbering system was changed to 6 digits to allow for more subsections in each division, for finer definition. In the new numbering system, the first two digits represent the division number. The next two digits indicate subsections of the division, and the two remaining digits signify the third-level sub-subsection numbers. The fourth level, if required, is a decimal and number added to the end of the last two digits. For example, the number 132013.04 indicates division 13, subsection 20, sub-subsection 13, and sub-sub-subsection 04.

Applications Guide

MasterFormat Groups, Subgroups, and Divisions

Procurement and Contracting Requirements Group

Division 00 – Procurement and Contracting Requirements
- Introductory Information
- Procurement Requirements
- Contracting Requirements

Specifications Group

General Requirements

Division 01 – General Requirements

Facility Construction Subgroup

Division 02 – Existing Conditions
Division 03 – Concrete
Division 04 – Masonry
Division 05 – Metals
Division 06 – Wood, Plastics, and Composites
Division 07 – Thermal and Moisture Protection
Division 08 – Openings
Division 09 – Finishes
Division 10 – Specialties
Division 11 – Equipment
Division 12 – Furnishings
Division 13 – Special Construction
Division 14 – Conveying Equipment
Division 15 – Reserved for Future Expansion
Division 16 – Reserved for Future Expansion
Division 17 – Reserved for Future Expansion
Division 18 – Reserved for Future Expansion
Division 19 – Reserved for Future Expansion

Facility Services Subgroup

Division 20 – Reserved for Future Expansion
Division 21 – Fire Suppression
Division 22 – Plumbing
Division 23 – Heating, Ventilating, and Air-Conditioning (HVAC)
Division 24 – Reserved for Future Expansion
Division 25 – Integrated Automation
Division 26 – Electrical
Division 27 – Communications
Division 28 – Electronic Safety and Security
Division 29 – Reserved for Future Expansion

Site and Infrastructure Subgroup

Division 30 – Reserved for Future Expansion
Division 31 – Earthwork
Division 32 – Exterior Improvements
Division 33 – Utilities
Division 34 – Transportation
Division 35 – Waterway and Marine Construction
Division 36 – Reserved for Future Expansion
Division 37 – Reserved for Future Expansion
Division 38 – Reserved for Future Expansion
Division 39 – Reserved for Future Expansion

Process Equipment Subgroup

Division 40 – Process Interconnections
Division 41 – Material Processing and Handling Equipment
Division 42 – Process Heating, Cooling, and Drying Equipment
Division 43 – Process Gas and Liquid Handling, Purification, and Storage Equipment
Division 44 – Pollution and Waste Control Equipment
Division 45 – Industry-Specific Manufacturing Equipment
Division 46 – Water and Wastewater Equipment
Division 47 – Reserved for Future Expansion
Division 48 – Electrical Power Generation
Division 49 – Reserved for Future Expansion

19 Introduction and Applications Guide

Figure 34 MasterFormat groups, subgroups, and divisions.

2.0.0 Section Review

1. Which type of architectural drawing would be most helpful for calculating the area of a room to determine the amount of wallboard needed?
   a. Site plan
   b. Reflected ceiling plan
   c. Floor plan
   d. Elevation drawing

2. Which drawings provide information on the size and placement of load-bearing elements as well as how load-bearing elements are connected to each other and other parts of the structure?
   a. Architectural
   b. Structural
   c. Mechanical
   d. Electrical

3. Drywall technicians need to be aware of the locations of certain mechanical, electrical, and plumbing components because _____.
   a. they may need to make passage for or work around MEP item
   b. they are directly involved in working on MEP systems
   c. OSHA requires it
   d. it helps them decide which type of drywall to use

4. True or False? Legally, written specifications take precedence over a set of drawings.
   a. True
   b. False

5. Specifications follow a standard format, usually the *MasterFormat*™, _____.
   a. as outlined in the building codes
   b. for convenience in writing and ease of reference
   c. because it is required by law
   d. as specified by ANSI

Module 45201 Review Questions

1. Architectural drawings are identified with the letter _____.
   a. *E*
   b. *A*
   c. *M*
   d. *P*

2. Structural drawings usually include _____ plans.
   a. HVACR
   b. riser
   c. piping
   d. foundation

3. A second-floor structural drawing would show the _____.
   a. placement of partition walls
   b. placement of electrical service drops
   c. floor layout
   d. configuration of load-bearing members

4. Mechanical drawings for HVACR systems are typically shown in a(n) _____ view.
   a. section
   b. plan
   c. elevation
   d. cutaway

5. Which type of drawings include the emergency lighting systems?
   a. Electrical
   b. Civil
   c. Mechanical
   d. Architectural

6. You can avoid missing details when reading commercial plans by using a _____.
   a. step-by-step process
   b. schedule
   c. local code
   d. plan handbook

7. True or False? On a commercial site plan, only existing contour lines are shown.
   a. True
   b. False

8. A feature on site plans that may be found only on commercial plans are _____.
   a. directional arrows
   b. exterior wall positioning data
   c. landscaping features
   d. grading details

9. For large buildings, the floor plan is often divided into sections by _____.
   a. colors
   b. coordinates
   c. grid lines
   d. patterns

10. Which of the following would include information about the dimensions, materials, finishes, and framing details of doors and windows?
    a. Door and frame schedule
    b. Window and door plans
    c. Door and frame key
    d. Window and door legend

11. An enlargement of an area on the floor plan is called a(n) _____.
    a. elevation
    b. schedule
    c. section
    d. detail

12. Exterior elevations are normally drawn to the same scale as the _____.
    a. floor plans
    b. site plans
    c. reflected ceiling plans
    d. MEP plans

13. What is the purpose of the elevation and section views in a set of plans?
    a. To provide the detail required for construction
    b. To provide an overall view of the proposed structure
    c. To show the height at which each section should be constructed
    d. To clarify ambiguous details on the structural plans

14. The placement of a column in a structural drawing will be shown by a(n) _____.
    a. circle with a smaller circle inside it
    b. circle with an X through it
    c. callout sequence number or mark
    d. open grid square

15. The structural engineer draws a framing plan for _____.
    a. each floor and the roof
    b. the plumbing, electrical, and HVACR systems
    c. landscaping and grading
    d. the foundation and exterior walls

16. The arrangement of beams, joists, and girders is shown on _____.
    a. elevations
    b. sections
    c. schedules
    d. framing plans

17. It is easier to read framing plans by _____.
    a. starting at the bottom and working upwards
    b. starting at the top and working downwards
    c. isolating the separate bays or spans between columns
    d. reading the foundation plans first

18. Concrete is used in building foundations because it has _____.
    a. low compressive strength
    b. high compressive strength
    c. low tensile strength
    d. high tensile strength

19. Why is steel reinforcement (rebar) used in concrete foundations?
    a. To decrease compressive strength
    b. To increase compressive strength
    c. To increase tensile strength
    d. To decrease tensile strength

20. For bonded post-tensioning systems used in elevated floors, the tendons are _____.
    a. threaded through plastic sleeves and tensioned after the concrete has cured
    b. tensioned before the concrete has been placed
    c. embedded in the concrete when the concrete is placed
    d. embedded in the concrete and tensioned off site at a manufacturing facility

21. Which type of plans typically include schedules that identify the different types of HVACR equipment?
    a. Structural
    b. Architectural
    c. Site
    d. Mechanical

22. A plumbing riser diagram shows _____.
    a. vertical pipes in the wall
    b. horizontal distance or pipe runs
    c. the hot- and cold-water supply lines
    d. the couplings, elbows, tees, and valves needed

23. There may be discrepancies between the written specifications and the working drawings because _____.
    a. the specifications were amended, but the drawings were not
    b. they refer to different parts of the project
    c. specifications are only used by the engineer and architect
    d. specifications are no longer needed once the drawings are prepared

24. Which of the following accompanies a plan set and provides detailed information on the products, parts, components, and installation methods for an entire project?
    a. Schedules
    b. As-builts
    c. Specifications
    d. Requests for information

Answers to odd-numbered review questions are found in the Module Review Answer Key found at the end of this book.

Answers to Section Review Questions

Answer	Section	Objective
Section One		
1. d	1.0.0	1
2. a	1.1.0	1a
3. a	1.1.0	1a
4. b	1.1.1	1a
Section Two		
1. c	2.1.2	2a
2. b	2.2.0	2b
3. a	2.3.0	2c
4. a	2.4.0	2d
5. b	2.4.2	2d

Steel Framing

Objectives

Successful completion of this module prepares you to do the following:

1. Describe the advantages of cold-formed steel as a framing material.
   a. Identify the factors that are considered when designing steel framing systems.
   b. Describe steel framing applications.
2. Describe tools, materials, and construction methods used in steel framing.
   a. Identify tools used for installing steel framing.
   b. Identify fasteners used for installing steel framing.
   c. Describe the purpose and types of steel framing connectors.
   d. Describe steel materials, how to identify them, and basic construction methods for installing them.

Performance Tasks

Under supervision, you should be able to do the following:

1. Build a section of curtain wall from shop drawings to include a window opening with headers, jambs, and sill.
2. Build headers (back-to-back, box, and L-header).
3. Lay out and install a steel stud structural wall with openings to include bracing and blocking.
4. Lay out and install a steel stud nonstructural wall with openings to include bracing and blocking.

Overview

In commercial and multifamily residential construction, it is common to use steel framing materials to frame walls and partitions. It is also becoming more common in single-family residential construction. As more designers and builders incorporate steel as a structural material, the need for skilled steel framing technicians continues to grow. This module describes the characteristics, applications, and installation of cold-formed steel framing.

Digital Resources for Drywall

Scan this code using the camera on your phone or mobile device to view the digital resources related to this craft.

1.0.0 Steel Framing

Performance Tasks

There are no Performance Tasks in this section.

Objective

Describe the advantages of cold-formed steel as a framing material.
  a. Identify the factors that are considered when designing steel framing systems.
  b. Describe steel framing applications.

Cold-formed steel: Steel products made from relatively thin metallic-coated coils, formed by processes carried out at or near ambient room temperature. These processes are different from hot-rolled steel, where uncoated steel's temperature is raised to near its melting point for forming into final shapes. Cold-forming processes typically include slitting, shearing, press braking, and roll forming.

Builders, especially those who are constructing the framework of a structure, look for certain key qualities in their building materials: strength, durability, consistency, cost-efficiency, sustainability, and functionality. **Cold-formed steel** (CFS) checks all of those boxes. CFS is becoming increasingly popular as a framing material due to the following characteristics:

- Steel studs and joists are strong and lightweight, with the highest strength-to-weight ratio of any building material.
- CFS does not warp, crack, rot, or split.
- Steel framing members are made with consistent dimensions, so assembly on the jobsite can be done rapidly.
- The exterior of steel members is uniform and provides a flat surface for sheathing or other material attachment. Steel is compatible with all sheathing materials except pressure-treated wood, which must be separated from steel to prevent corrosion.
- Steel framing does not contribute combustible material to feed a fire. Its lower combustion rating means lower insurance costs for builders and owners.
- Steel can be engineered to meet the toughest wind and seismic ratings specified by building codes.

To understand what gives cold-formed steel its qualities, it helps to be familiar with its production process. Cold-formed steel is not actually produced at cold temperatures. The process of making both hot-formed (usually referred to as hot-rolled) and cold-formed steel begins with very hot raw, molten metal. The term *cold-formed steel* is used in comparison to hot-rolled steel, which is formed by heating the metal and then shaping or molding it before it recrystallizes. Cold-formed steel is compressed and shaped at room temperature, after it has already recrystallized. Although re-shaping steel after it has cooled is trickier than working with hot steel, it results in a more durable and consistent product because, unlike hot-rolled steel, it doesn't continue changing shape (shrinking) after it is formed.

NOTE

Appendix 45202A contains a list of terms commonly used in conjunction with cold-formed steel framing.

Did You Know?

Steel Is Sustainable

In addition to its good structural features, steel has some important environmental qualities. Industry-wide, steel has an average of 67 percent recycled content, making it the world's most recycled material. In the United States alone, almost 70 million tons of steel per year are recycled or exported for recycling. Because of its light weight, cold-formed steel impacts the environment less than heavier materials, which produce more emissions when being transported or lifted. Furthermore, CFS structures are durable and require little maintenance over time, which reduces environmental impacts from construction activities. Cold-formed steel framing supports industry efforts to promote sustainable construction.

Source: JohnDWilliams/Getty Images

As the demand for more cost-efficiency and sustainability in construction grows, builders and developers are specifying both structural and nonstructural applications of cold-formed steel. CFS framing techniques can be used to construct load-bearing and nonbearing walls, exterior and interior wall sections, floor decking, roof trusses, and a variety of other features. While cold-formed steel was used as a framing material without regulation as early as the 1850s, recognition and acceptance of CFS in building codes is leading to a better understanding of its capabilities and benefits. Low and mid-rise structures using cold-formed steel as the main structural component have been successfully built throughout the United States, creating highly durable and safe structures while lowering construction costs. In order to take full advantage of steel's structural strength, it is essential for framing technicians to understand the core concepts of structural engineering.

1.1.0 Structural Engineering

Structures are designed to withstand the forces, or stressors, acting upon them. When a building has too little structural support, it can collapse under stress. When a building has too much structural support, it wastes building materials, labor, and time. Engineers study the internal and external stressors to which a building will be subjected, and then they design the structure to withstand them. As a framing technician, it is important to understand these factors and the importance of implementing the design exactly according to the plans and specifications.

The following stressors, or loads (*Figure 1*) are considered when a structure is designed:

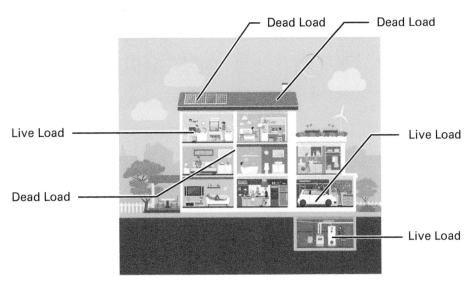

Figure 1 Building stressors.
Source: elenabs/Getty Images

- *Dead loads* — The weight of the building materials on the structure. Any component that is a permanent part of the structure, such as the roof, beams, columns, windows, floors, and even rooftop air conditioners, are considered dead loads. In addition, any force applied to the building during construction is a dead load. For example, a crane or hoist that is used on the roof during the construction of a building would be a dead load.
- *Live loads* — Weight that results from using the building when it is finished and occupied. Examples of live loads include people, furniture, appliances, pets, and everything else occupying the space inside the structure.

- *Other loads* — In addition to dead and live loads, buildings are subjected to stress from wind, rain, snow, natural disasters, and pressure from groundwater.

One type of force that is of special importance to tall structures is wind. Wind causes tall buildings to move, similar to the way a tall tree sways in the wind. This movement is called *horizontal deflection*, also known as **lateral** drift. All tall buildings are designed to move with the wind. A wind blowing at 110 miles an hour will move the Empire State Building about 1.48 inches at the top.

Designers must address all the various types of stressors when designing buildings. Whether the building is constructed of wood framing or steel framing, the structural factors are the same. The building must support its dead load and its live load, as well as other loads, such as snow, rain, wind, and weather events.

Buildings are constructed so that the weight from all loads is transferred through structural members to the foundation until all of the weight is transferred to the ground. This design transfers the weight of the roof to the wall posts, then to the floors and foundation. Finally, all the weight of the building is transferred to the ground. *Figure 2* illustrates the path that the weight of gravity takes as it pushes down on a structure.

Lateral: Running side to side; horizontal.

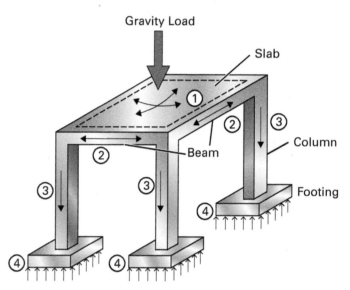

Figure 2 Load path through a structure.

Not all wall studs are load-bearing. Some walls only need to support their own weight. Interior and exterior walls may be either load-bearing or nonbearing. Load-bearing walls need to be strong, and they require a great deal of support. Nonbearing walls only need to support themselves. A **curtain wall** is an exterior wall that is designed to support its own weight and to resist wind pressures, but it is not designed to support any other load.

Because door and window openings cause a break in the wall stud pattern, they need to be reinforced on the top and on each side to make up for the missing studs. Reinforcement of door and window openings includes **headers**, jack studs, king studs, cripple studs, and sill plates. Headers are placed horizontally over the door and window openings. To increase structural support, *jack studs* (shorter, vertical support studs) are butted against and secured to full length wall studs called *king studs*. This combination of king stud and jack stud is referred to as a *jamb stud* and is placed on each side of the opening to support the headers. Beneath the window openings, *cripple studs* are placed vertically underneath the horizontal framing member called the *sill plate*. Window and door openings may be purchased as pre-engineered items, or they may be built on the jobsite. Steel framing members may be fastened together by screws, welding, or cinching according to the project specifications.

Curtain wall: A light, nonbearing exterior wall attached to the concrete or steel structure of a building. It primarily resists wind loads and supports only its own weight and cladding weight.

Headers: Horizontal structural framing members used over floor, roof, or wall openings to transfer loads around the opening to supporting structural framing members.

Cold-Formed Steel Identification

The use of cold-formed steel members in building construction began around 1850. In North America, however, steel members were not widely used until 1946, when the American Iron and Steel Institute (AISI) Specification was first published. This design standard was primarily based on research sponsored by AISI at Cornell University. Subsequent revisions to the document reflected technical developments and ultimately led to the publishing of the North American Specification for the Design of Cold-Formed Steel Structural Members, referred to as *AISI S100*. These and other CFS specification documents are available as free downloads at www.cfsei.org (Cold-Formed Steel Engineers Institute). AISI, along with the American National Standards Institute (ANSI), American Society of Testing and Materials International (ASTM International), and the International Code Council (ICC), govern the design, manufacture, and use of cold-formed steel and framing. All framing members carry a product identification to comply with the manufacturer's name, minimum sheet steel thickness, the coating designation, and the minimum yield strength.

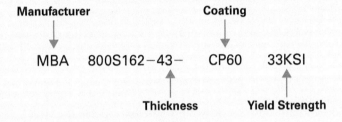

1.2.0 Steel Framing Applications

Historically, cold-formed steel framing has been widely accepted for use in non-structural applications, also referred to as nonbearing or partition walls. Today, CFS is commonly used for structural walls, floor assemblies, ceiling framework systems, and roof assemblies as well.

1.2.1 Framing Structural (Load-Bearing) Walls

A load-bearing or structural wall supports the weight above it by transferring the weight through itself to the floor or foundation below. Whether constructed of wood or steel, structural walls have the same basic components. *Figure 3* illustrates the fundamental parts of a structural wall.

Structural walls support the weight of the building and protect occupants from wind loads and other forces of nature. A basic, cold-formed structural steel wall includes structural studs, **tracks**, fasteners, bracing, and bridging. To add support to structural walls, headers for window and door openings are constructed using structural studs, joists, or proprietary cold-formed shapes.

Tracks: Framing members consisting of only a web and two flanges. Track web depth measurements are taken to the inside of the flanges.

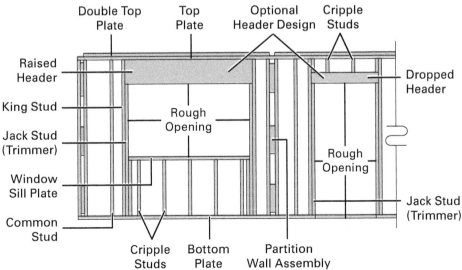

Figure 3 Parts of a structural wall.

Mils: Units of measurement equal to $\frac{1}{1,000}$ of an inch.

Prefabricated Panel Building Construction

Today, a number of companies are offering services for prefabricating panelized cold-formed steel exterior and interior structural walls, interior partitions, floors, and roofs in accordance with architectural plans. The prefabrication is accomplished off-site in a factory and the panelized sections are shipped assembled to the construction site. Some companies provide prefabrication of the panelized sections on-site in mobile facilities. In either case, stick-fabrication of the building is virtually eliminated.

NOTE

Steel studs are typically spaced 16 inches apart from the center of one to the next, also known as 16 inches *on center*. You may see plans that call for them to be spaced at 24 inches on center, which is typically done for temporary walls.

CAUTION

If the dimensions are not provided, seek clarification. Never attempt to scale the dimensions from a drawing. A small mathematical error could result in dangerous and costly structural problems.

Wall framing members for load-bearing walls have a material base steel thickness of 33 to 118 mils, with a minimum metallic zinc coating of G60. Typically, structural wall studs are produced in sizes from $2\frac{1}{2}$ to 8 inches in width, with flanges ranging from $1\frac{3}{8}$ to $2\frac{1}{2}$ inches or more. The return lips (edge stiffeners) depend on the flange size. Structural track is sized to accommodate the web depth of wall studs, and the flanges are typically sized $1\frac{1}{4}$ to 3 inches. The selection of a particular member will depend on its intended use. A number of proprietary and nonproprietary accessories are also available for structural wall assembly, including cold-rolled channel for bridging, as well as clips, angles, and straps.

Layout

Taking the time to plan the placement of wall studs before installing them is critical because they must align properly with other parts of the structure such as the floor and roof joists. Additionally, drywall technicians will need the studs to be correctly spaced and plumbed.

Place the top and bottom track members on the straight edge of the panel table. They should be arranged with the webs next to each other, clamped together temporarily. They should fit tightly against the edge of the end stop (the straight edge at the end of the wall). Mark the layout of the wall studs on the flanges of the top and bottom track, starting with a wall stud at the end of the wall. Use highly visible ink, such as a black felt-tip marker. Place a line at the web location and an X on the side of the line to indicate the stud flanges.

Mark the next stud location to match the first truss or roof rafter location from the end wall. Continue marking every 16 inches (or 24 inches, depending on the layout) for the full length of the wall. Where the exterior corner walls intersect, the wall that runs to the edge of the foundation has an extra stud. This stud is 3 inches from the end and acts as a backer to screw to the shorter intersecting wall.

Next, identify the rough openings in the walls. Check the architectural drawings to find door and window locations.

Mark the location for the center of the openings on the top and bottom tracks. Use a red felt-tip marker to distinguish these marks from the layout marks. Check the door and window sizes on the drawings and verify rough openings with actual window sizes. Add 12 inches to the width of the window openings to allow for two jack studs on each side of the header with room to spare.

Using a tape measure, center the dimensions over the red marks on the track. Mark each end of the tape measure. This shows the location of the webs of the king studs. Put the X on the side of the mark away from the window. The webs of the king studs will be on the rough opening side. This dimension is standardized to simplify header ordering and assembly. Two king studs may not be required at every opening, and framers may vary the length of the headers. However, standardizing header lengths helps to simplify cut lists.

All load-bearing studs must be aligned with the trusses, joists, or rafters above or below the wall. Because these walls carry loads, it is important to fit each stud tightly into the track member to allow the stud to properly carry the *axial* (downward) load from above. The top and bottom wall tracks must be of equal or greater thickness than the studs. Panelized walls may be constructed on a concrete slab, floor deck, or panel table. For all systems of wall construction, make sure that the surface on which the wall is built is level. Walls must not be out of plumb more than $\frac{1}{8}$ inch for every 10 feet.

Walls may be framed in two ways. Full-length walls may be framed up to 40 feet, depending on available personnel at the jobsite. Be aware that a longer wall tends to twist and could get damaged if the crew is not large enough to keep the wall straight. Shorter wall sections need more plumbing and alignment, but they are easier to assemble and may be built and spliced together. The track will also need to be spliced (*Figure 4*).

Wall Assembly

Before beginning assembly of a structural wall, make sure that the foundation or the bearing surface is free of all defects. The bearing surface should be uniform, with a maximum $\frac{1}{8}$-inch gap between it and the track.

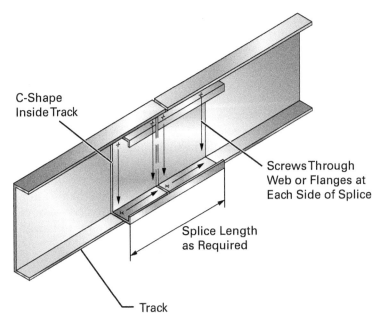

Figure 4 Track splicing.

After layout has been completed, separate the top and bottom tracks, and install a wall stud at each end of the wall between the top and bottom tracks. Clamp the stud flanges to the track flanges with locking C-clamps at each end. Tap the track on one end with a hammer to seat the studs as tightly as possible in the track. Fitting each stud tightly and perpendicular to the track keeps the wall straight. Screw one low-profile No. 10 screw through the flange of the track into the flange of the stud on both sides of the stud. If an elevated panel table is used, the framer may be able to install the screw (from underneath) on the other flange as well. If not, once all the studs and headers are in place and all screws are installed on one side, flip the wall over to install the screws on the other side.

In drywall (nonstructural) framing, the only stud-to-track screw connections required are at wall corners and at jamb studs. However, framers often use more screws to aid with the stability of the framing. Check local codes. For load-bearing walls, screws must be installed on each flange on both sides of the wall. This keeps the studs from twisting and provides the proper connection for in-line framing. Continue twisting the studs into the track. Install all the studs the same way, with the open side of the **C-shape** (*Figure 5*) facing in the same

C-shape: A cold-formed steel shape used for structural and nonstructural framing members consisting of a web, two flanges, and two lips (edge stiffeners).

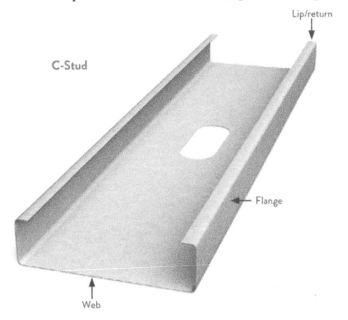

Figure 5 C-shape (also known as C-section, C-stud, and C-channel).
Source: Image provided by ClarkDietrich

direction and toward the start of the layout. Align the punchouts in the studs to provide straight runs for the plumbers and electricians. The studs should be aligned so that they all face in the same direction on parallel load-bearing walls.

Install the king studs at the rough openings with the hard side of the stud facing the rough opening and punchouts aligned. Do not install studs at the markings between the king studs. These markings will be for the cripple studs when the rough openings are framed. Continue down the length of the wall until all the studs are twisted and screwed into place. Do not remove the wall panel from the table until the headers and rough openings have been completed.

During wall assembly, it is important to seat the studs into the track as tightly as possible. Studs must be seated with a gap of no more than $\frac{1}{8}$" for structural (load-bearing) framing, and no more than $\frac{1}{4}$" for nonstructural or curtain wall framing. This ensures that the building loads are transferred through the studs, and not through the fastener (*Figure 6*).

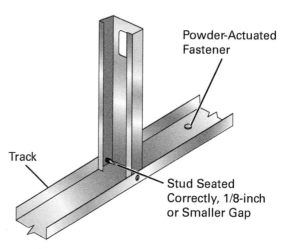

Powder-Actuated Fastener

Track

Stud Seated Correctly, 1/8-inch or Smaller Gap

Figure 6 Proper stud seating.

Structural walls may be constructed using any of the following methods:

- *Stick building* — Walls are framed in place, one stud at a time.
- **Panelization** — Walls are assembled and then moved to their installation location.
- *Pre-manufactured method* — This assembly typically increases the size and spacing of structural steel members. In some cases, the spacing can be as much as 4 or 6 feet on center.

Panelization: The process of assembling steel-framed walls, joists, or trusses before they are installed in a structure. Roll-formers normally cut studs to within $\frac{1}{8}$-inch tolerance. This helps the framer to consistently create straight walls on the panel table that are easy to install in the field.

Did You Know?

Steel Framing Is Customizable

Prefabricated metal framing systems can be customized to create unique design elements. All or part of a structure's frame can be engineered to specifications and assembled off-site. Customized studs, wall panels and systems, clips and fasteners, floor joists, and roof trusses can be ordered and installed as separate pieces on-site. Customization of steel framing materials gives designers the ability to create eye-catching, contemporary architecture while taking advantage of the strength and durability of steel.

Screws are commonly used to attach track to cold-formed steel floor or roof assemblies, but welding may be used when specified. Powder-actuated fasteners may be required when the track connects to structural steel framing members or concrete. Expansion bolts are used at jamb locations and corners, while expansion anchors are typically used at **shear wall** locations.

Wall Installation

If anchor bolts are used in the foundation, measure their locations and place holes in the bottom track of the wall panels so that the walls will fit over the bolts. If strap anchors or other kinds of anchors are used, this step will not be necessary. Place temporary bracing material near the foundation in preparation for the wall to be raised. Any stud material at the jobsite, preferably 12 feet long, may be used. Caulk the concrete foundation with weatherproof caulking material and use foam, closed-cell sill sealer.

Move the wall panel and set the bottom track on the foundation. Position it over the anchor bolt locations, which may need to be adjusted, and tilt the wall up. Clamp the temporary brace material to the wall studs in two or three locations, depending on the length of the wall. Temporary bracing is often placed on the inside of the structure. It can also be placed alternating on both sides of a wall, or only on the outside. Before removing the clamps, secure a No. 10 hex head screw through the brace into the stud. Install a brace every 8 to 12 feet along the wall. Secure the bottom of the brace with a stake driven into the ground or other solid surface. Screw through the stud into the stake to hold it in place. Repeat this process with all wall panels until all load-bearing walls are standing.

WARNING!

Before tilting up a wall assembly, make sure that the braces do not hang past the wall edges. Snagging on the braces is very dangerous and could result in injury.

Framers may construct walls using studs that span continuously from the foundation to the roof (balloon framing) or they may construct one floor at a time using studs that rest on the floor joist between each floor (platform framing). *Figure 7* illustrates the difference between balloon and platform framing. When balloon framing is used, installers must follow the shop drawings exactly because the studs are pre-engineered to meet specifications.

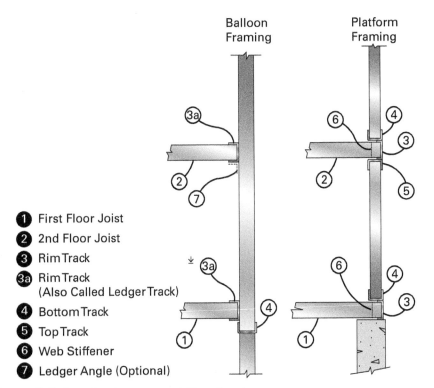

1. First Floor Joist
2. 2nd Floor Joist
3. Rim Track
3a. Rim Track (Also Called Ledger Track)
4. Bottom Track
5. Top Track
6. Web Stiffener
7. Ledger Angle (Optional)

Figure 7 Balloon framing versus platform framing.

Shear Walls

Lateral (shear) loads are live loads that act on a building horizontally. Unlike gravity, which is a vertical force pushing down on a structure, lateral loads come from the side. Wind is the primary example. Shear wall construction is especially important in hurricane- and earthquake-prone areas. Although earthquakes may not seem like lateral forces because they are generated from the ground, the side-to-side shaking movement produced by the earthquake places lateral loads on a structure. Framers construct shear walls to resist everyday horizontal forces as well as those that might be experienced during a natural disaster. Shear walls are typically used in conjunction with other lateral support structures such as floor and roof/ceiling **diaphragms** to transfer lateral loads to the foundation of a building. Shear walls can be created by attaching gypsum board, wood structural sheathing, steel decking, X-bracing (*Figure 8*) or other materials to the floor, wall, ceiling, or roof framing. In addition to field-fabricated shear wall systems, there are now several proprietary, high-strength pre-engineered systems that may help shorten construction time (*Figure 9*).

Diaphragms: Floor, ceiling, or roof assemblies designed to resist in-plane forces such as wind or seismic loads.

Figure 8 X-bracing.
Source: Don Wheeler

Figure 9 High-strength, pre-engineered shear wall system.
Source: Don Wheeler

NOTE

All concealed cavities such as headers in exterior walls must be pre-insulated before they are installed.

The wall design for any steel-framed structure should provide for the placement of shear walls. In multistory construction, proper alignment (stacking and load path) is also required to ensure that there is adequate shear transfer between roof or floor diaphragms and shear walls or other assemblies such as bar joists, long-span deck, and wood framing. Shear walls are usually connected to the foundation using proprietary or engineered connectors and hold-downs.

Header Assembly

Load-bearing headers are typically boxed, unpunched C-shapes that are capped on the top and bottom with track sections. C-shapes are a cold-formed steel shape used for structural and nonstructural framing members. C-shapes consist of a web, two flanges, and two lips (edge stiffeners). They must be engineered for bending and shear strength, along with web crippling (crushing) at the locations of the loads from above. The two types of headers that are most commonly built from standard C-shapes is the box header (*Figure 10*) and the back-to-back header (*Figure 11*). A third type, called an L-header (*Figure 12*), is a steel angle in the shape of an L. Prefabricated headers may also be ordered from manufacturers. Selection of the header type depends on loads and applications. L-headers are being used more frequently in low-rise and multifamily construction because they use fewer fasteners and less material.

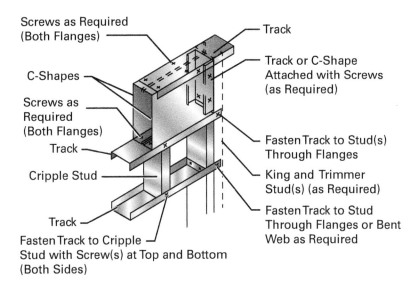

Figure 10 Box header.

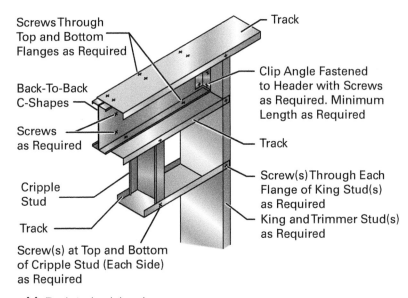

Figure 11 Back-to-back header.

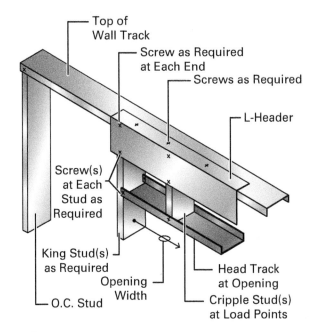

Figure 12 L-header.

To install the box header, loosen one of the king studs at the top of the rough opening, and install the header with the open end up, by inserting it into the top track. This is usually a tight fit. Clamp the header at one end with a bar clamp. Apply pressure, working from one side of the header to the other, using the clamp to tightly push the header into the top track. Make sure that the header is fitted tightly into the track before screwing it in place. Reposition the king studs that were loosened and screw them back into place.

Back-to-back headers are formed by placing two C-shapes with the webs of the members touching each other. They are positioned in the top track of the wall and finished just like a box header.

The L-header (*Figure 12*) consists of one or two angle pieces that fit over the top track. This design saves labor because no special fabricating is needed, and the number of screws is reduced. The L-shape itself spans the opening for the header.

Once the header has been installed in the top track, the remainder of the rough opening may be framed (*Figure 13*).

Figure 13 Completed rough opening.
Source: Printed with permission from the American Iron and Steel Institute

Mark the jack studs for the top and bottom of the window. Cut two track pieces to fit between the jack studs to make up the head and sill pieces. These pieces should be cut 2 inches longer than the distance between the jack studs so the flanges can be clipped 1 inch on each side.

Clip the flanges and bend the web down toward the flanges. Set the head and sill pieces in the opening, keeping the hard side of the track towards the opening. Screw the tab at the flange of each piece into the jack studs with one No. 10 screw at each tab. The tabs on the webs of the header track should also be screwed into the jack studs with two No. 10 screws.

The cripple studs should be cut to fit between the header track and head pieces, as well as between the sill piece and bottom track. The cripples should maintain the spacing layout (16 or 24 inches on center) for ease in installing gypsum board and sheathing. Screw the cripple studs into place with a No. 10 screw at each track flange on both sides.

Door openings do not require a bottom sill. However, the bottom track should run continuously at the bottom of the door to hold the wall together temporarily. The track can be cut out after the wall is plumb, level, and permanently braced.

NOTE

Box and back-to-back headers are typically designed as unpunched. Manufacturers can easily produce unpunched (solid web) material as a special order.

NOTE

After header assembly, the window opening on the drawing should be rechecked to verify the position and height of the window in the wall.

Jambs

Load-bearing jambs require a minimum of two studs on each side of the framed opening (one jack stud plus one king). More studs will be needed if they are required by the building code or by engineering analysis. The studs must be fastened together to act as one member, either by capping them with track and screw fastenings or by stitch welding.

Sills

Sill members are usually single-track sections that are the same width as the wall stud used to frame an opening. They are clipped and screwed to the jamb studs. In cases where the allowable lateral load is exceeded, sills must be constructed with multiple track sections.

Bracing Steel Walls

Bracing prevents a cold-formed steel framing member from twisting and buckling. In wall construction, there are several common methods for bracing, and all have different applications and purposes. *Figure 14*, *Figure 15*, and *Figure 16* show examples of typical types of bracing.

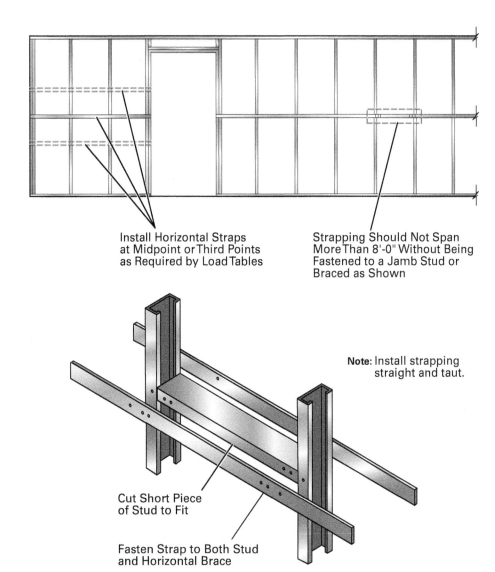

Install Horizontal Straps at Midpoint or Third Points as Required by Load Tables

Strapping Should Not Span More Than 8'-0" Without Being Fastened to a Jamb Stud or Braced as Shown

Note: Install strapping straight and taut.

Cut Short Piece of Stud to Fit

Fasten Strap to Both Stud and Horizontal Brace

Figure 14 Lateral strapping for stud walls.

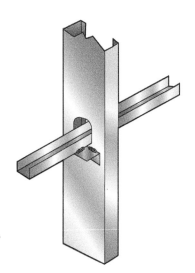

Figure 15 Bridging with cold-rolled channel (CRC).

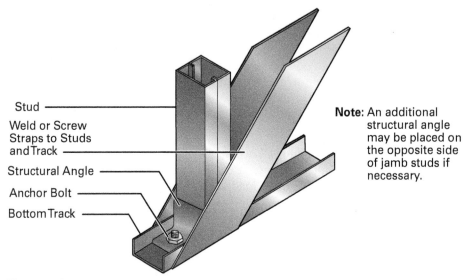

Stud

Weld or Screw
Straps to Studs
and Track

Structural Angle

Anchor Bolt

Bottom Track

Note: An additional
structural angle
may be placed on
the opposite side
of jamb studs if
necessary.

Figure 16 Diagonal strapping for shear walls.

Intermediate stud bracing — This bracing is used when gypsum board or structural sheathing is not applied to both sides of load-bearing walls, such as garage walls (*Figure 17*).

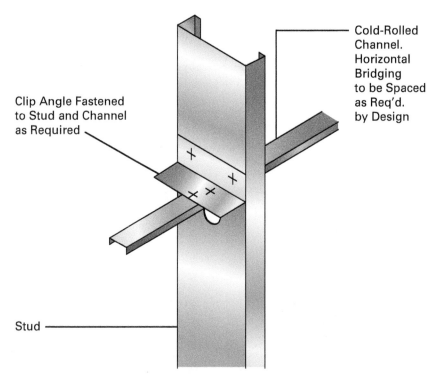

Clip Angle Fastened
to Stud and Channel
as Required

Cold-Rolled
Channel.
Horizontal
Bridging
to be Spaced
as Req'd.
by Design

Stud

Figure 17 Stud bracing using cold-rolled channel.

Racking: Being forced out of plumb by wind or seismic forces.

Shear wall bracing — There are two ways of applying shear wall bracing: structural sheathing (*Figure 18*) and X-bracing (*Figure 19*). Depending on the design, structural sheathing Type II plywood or OSB may be adequate to keep the wall from **racking** if there are not excessive openings in the wall. For structural sheathing to be effective, it should be installed with the long dimension parallel to the stud framing (vertical orientation). The plywood may be secured to the wall while panelizing, or after the wall is plumb and level. X-bracing is another way to obtain shear strength when structural sheathing is not used. X-braces are diagonal steel straps attached to the walls with screws or welded connections. This bracing must be designed by an engineer, and the straps must

Figure 18 Wood structural sheathing.
Source: Printed with permission from the American Iron and Steel Institute

Figure 19 Shear wall X-bracing system.
Source: Image provided by ClarkDietrich

be inspected for the correct number of fasteners. Do not tighten the straps until the walls are plumbed and aligned.

Temporary bracing — There are two types of basic temporary bracing: one for panelized walls and one for installed walls. After a straight wall is constructed on a panel table, installing plywood or temporary bracing prevents racking when the wall is removed from the table. Before taking the wall off the table, check for squareness by diagonally measuring the panel. When both diagonals are equal, the wall is square. Adjust if necessary. Lay extra studs or truss material across the wall diagonally, and screw the bracing to the wall studs, especially at door openings where the bottom track is weak. Leave the bracing on the wall until the wall is installed and permanently braced. These precautions will help provide straight walls that are ready for final installation.

Installation of steel-framed walls requires adequate temporary bracing to resist loads during construction until permanent bracing can be installed (*Figure 20*).

Figure 20 Temporary bracing.
Source: Don Wheeler

1.2.2 Nonstructural (Nonbearing) Wall Framing

Steel framers also construct nonstructural, or nonbearing, wall systems. Nonbearing walls do not support the weight of the floors above them and are not responsible for transferring structural load. Most nonbearing walls function as space partitions within the exterior walls of a building.

Similar to structural wall systems, nonstructural walls are built using studs, track, and accessories. The primary differences are the characteristics of the materials and the application of connectors and accessories.

Nonstructural framing members have a thickness ranging from 15 to 33 mils, compared with a minimum thickness of 33 mils for structural studs (*Table 1*). In addition, the minimum stud flange dimension for nonstructural framing members is $1\frac{1}{4}$ inches, and the minimum return (stiffening) lip dimension is $\frac{3}{16}$ inch, compared with a $1\frac{5}{8}$-inch flange and $\frac{1}{2}$-inch lip for most structural studs. Also, nonstructural members typically have a G40 galvanized coating weight, compared with G60 or higher for structural studs. The rules for nonstructural framing are different from those for structural walls. These rules may be found in the gypsum specifications, rather than in the building codes.

TABLE 1 Nonstructural Versus Structural Steel Framing

Nonstructural	Structural
Framing members have a thickness ranging from 15 to 33 mils.	Framing members must have a base thickness of at least 33 mils.
Minimum stud flange dimension is $1\frac{1}{4}$ inch.	Minimum stud flange dimension is $1\frac{5}{8}$ inch.
Minimum return lip dimension is $\frac{3}{16}$ inch.	Minimum return lip dimension is $\frac{1}{2}$ inch.
Usually have G40 galvanized coating weight.	Have G60 or higher galvanized coating weight.
May have up to a $\frac{1}{4}$-inch stud-to-track gap.	May have up to a $\frac{1}{8}$-inch stud-to-track gap.

Because nonstructural studs are not intended to carry any loads, accessories used in this application are used for architectural features within the building, including curves and arches. No connection is required between the stud and track for nonbearing walls, except for studs that are adjacent to window and door openings and studs at partition intersections or corners.

Steel Curtain Walls

Since their introduction in the early 1900s, metal and glass curtain wall systems have become very popular in the architectural design of contemporary structures. Unlike interior partitions, wind-bearing curtain walls resist loads from exterior wind pressures that, in some cases, exceed 60 pounds per square foot. Cold-formed steel curtain walls are made up of various components, which may include the following:

- *Angles* — Angles are used to connect framing members within the curtain wall system.

- *Clip angle* — A steel angle, generally 3 to 12 inches long, that makes the transition between a framing member and the component supporting it. These angles are used to connect two framing members.

- *Continuous angle* — A steel angle that makes the transition between a stud curtain wall and the primary frame. The angle is typically of hot-rolled thickness ($\frac{3}{16}$ to $\frac{3}{8}$ inch); however, thinner materials can be used if the span and load requirements are relatively small.

- *Diagonal brace (or kicker)* — A sloping brace used to provide lateral support to a curtain wall assembly. When installed horizontally, a diagonal brace is referred to as a *strut*.

- *Embed* — A hot rolled steel plate or angle, reinforced with shear studs or steel rebar, which is cast into a concrete floor or beam. Embeds allow for the welded attachment of steel supports.
- *Girts* — Horizontal structural members that support wall panels and are primarily subject to bending under horizontal loads, such as wind load.
- *Slide clip* — A connection device that permits the primary frame to which a stud attaches to move axially, while it braces the stud against lateral forces.
- *Slip track* — A track section used in **for-fill** curtain wall applications (*Figure 21*). Slip tracks accommodate vertical movements of a primary frame (normally $\frac{1}{4}$-inch to $\frac{3}{4}$-inch), while bracing the wall against lateral forces. Slip tracks may also be specified at the top of interior drywall partitions.

For-fill: A wall to be filled with some type of material, such as concrete or insulation.

Figure 21 Slotted slip track.
Source: Don Wheeler

The term *curtain wall* is used to distinguish this system from load-bearing framing. As described earlier in this section, load-bearing framing requires that the wall members carry the weight of the structure that is above them. With curtain wall framing, the structure is usually already in place, and the wall framing is filled-in between the floor slabs. The stud-to-track gap distance for curtain wall is no more than $\frac{1}{4}$ inch unless it is otherwise specified in an approved design. This is different from load-bearing construction, which permits only a $\frac{1}{8}$-inch gap. The only gravity loads that curtain wall framing typically carry are the weight of any cladding or finish materials attached to it.

The following are four main methods for building assemblies in curtain wall construction:

- *Infill* — This term describes applications in which studs are only one story tall, spanning from floor to floor of a structure. Infill framing requires less stud material. Connections are often easier to make because low-cost powder-actuated pins can be used in many applications. In addition, the spans are often shorter than bypass conditions, so thinner steel or wider spacing can be used. The disadvantage of infill framing is that more track material is needed, and wall sections can be difficult to panelize.

CAUTION

For curtain wall installation, the use of materials formed from steel measuring less than 33 mils is not permitted per code.

- *Panelization* — Large sections or entire wall assemblies can be assembled and moved to the installation location. This method may be used if field measurements are made after the floor systems are in place, or if a slip connector or telescoping stud system is used.

- *Bypass framing (balloon framing)* — This technique allows a single stud to be used for framing two or more floors, and it requires less track material than other types of framing. The multiple spans can reduce moment stresses in members, allowing for more widely spaced or thinner framing members. In addition, this method makes it easier to pre-panelize large sections, including multistory panels. On the other hand, some connections can be more difficult to make, such as bracing and support connections behind columns and **spandrel beams**. Depending on the conditions, double connections may need to be made. This requires that clips or slip connectors are attached first to the structure and then again to the stud. Connection difficulties may sometimes be avoided by the use of connectors that friction-fit inside the stud, thus reducing connection time. If slab or other structural elements extend into the stud cavity, they may need to be chipped away, or stud framing may need to be altered to correctly install the bypass framing.

- *Stacked wall framing* — In stacked framing, the studs are aligned from floor to floor so that the structural load is transferred directly, eliminating the need for double top plates and reducing the amount of framing material needed. This method permits bypass framing while isolating slip connections to one- or two-story segments. As a result, multistory panels and pre-finished panels can be fabricated and installed, including insulation. However, to prevent water infiltration, this approach requires special detailing at slip connections and exterior finishes. Because the entire panel weight goes to fixed connections, usually at every other floor, these fixed connections need to be strengthened to carry the added dead load of the taller panels. Also, some connections are more difficult to reach, such as bracing and support connections behind columns and spandrel beams.

Spandrel beams: Structural support elements at the outer edge of a building or edge of a slab.

The most common cold-formed steel curtain wall shape is the C-stud. Due to its shape and geometric properties, a C-stud tends to rotate under lateral load. Non-braced studs may also move out of plane. This is known as *torsional-flexural buckling*. Mechanical bridging and/or sheathing materials can restrain the flanges and prevent this condition. When using discreet bracing rather than sheathing, decreasing the spacing of the bridging typically increases the member capacities. Increasing the spacing decreases capacities.

In a typical stud-framed exterior curtain wall, member deflections are the primary structural issue. Horizontal deflection was explained in *Section 1.1.0* as the amount of lateral movement caused by wind. *Deflection* also refers to the amount of movement of a structural component due to the force of any load placed on it. Framing systems must meet certain deflection limits which are most often determined by the architectural finishes and are set forth in the project specifications.

A wide range of finish systems may be used to finish steel stud curtain wall frames, including the following:

- Brick veneer
- Split-faced block veneer
- Tile or thin-cast brick
- Exterior insulation finish systems (EIFS)
- Glass fiber reinforced concrete (GFRC)
- Metal panel
- Modified portland cement (stucco)
- Fiber-cement board or siding
- Dimensional stone, such as granite or limestone
- Wood siding and many other finish systems

Curved Walls

Some partitions that might be difficult to construct with wood framing members can easily be built with steel framing members. A curved wall is an example. Note that the wall forms a quarter circle. Constructing a wall of this curvature with wood framing members would take many hours. Using steel framing members, the wall could be built in less than an hour. A plywood template may be required for more complex curved walls.

The bend can be achieved using several methods. Curved walls can be framed out of steel using curved track for partitions or exterior load-bearing walls. The track can be bent at the jobsite by slitting the flanges. Ordering curved track from specialty companies also speeds up the construction of a curved wall.

Track may also be ordered to a specified curvature. Some specialty companies use machinery to bend the track without slitting the flanges; others produce a flexible track that can be ordered and formed on the jobsite based on the desired effect. This provides a clean, neatly bent track to an exact curvature. Wall track is bent around the flanges (*Figure 22*).

Figure 22 Example of a curved track.
Source: alacatr/Getty Images

Other Nonstructural Walls

In addition to curtain walls, steel framers construct other types of nonstructural walls. Examples include a partial partition, or a chase wall designed to conceal wiring or plumbing. The following nonbearing wall assemblies have specialized functions, such as stopping the spread of fire and suppressing sound:

- *Chase walls* — It is common in commercial and residential construction to build chase walls (*Figure 23*) to conceal plumbing and other utilities. For over-sized utilities or acoustical requirements, two separate walls with a void in between may be framed in advance. Cross-bracing between the two walls may be required.

- *Fire-resistance-rated assemblies* — Building codes require that certain partitions provide fire-resistance-rated separation of interior spaces, including a specific rating that states how long the partition or assembly would last under

Figure 23 Chase wall for plumbing.
Source: imagenta/Shutterstock

the design load before being penetrated by a fire. Approved assemblies must be constructed precisely as described in the directories published by the rating agencies.

- *Area separation walls (zone walls)* — This term refers to a specific group of rated walls, designated by code requirements, which are designed to permit structural failure on one side of the wall while still providing fire protection.

- *Head-of-wall conditions* — Some building codes require that wall assemblies accommodate horizontal and vertical movement while still providing a fire-resistive barrier. Proprietary and nonproprietary devices and assemblies have been designed and tested to meet code requirements.

- *Shaftwall systems* — These structures are special types of rated wall systems that may be constructed from one side only. Higher sound and fire-resistance ratings may be achieved by attaching additional layers of gypsum board to the outer or inner face. A one-inch core board is specifically made for this purpose. Horizontal shaftwalls may also be used for a soffit or **plenum** where ratings are required.

Plenum: A confined space, such as the area between a suspended ceiling and an overhead deck, that is used as a return for a heating, cooling, and ventilation system.

Furring

Design requirements may specify a space between gypsum board materials and framing studs. This is accomplished through a technique called *furring*, which uses hat-shaped or Z-shaped steel members to create a standoff (additional space) between the frame and the gypsum board. Furring may be used for a variety of purposes, including the following:

- Providing additional space for insulation
- Allowing out-of-plane walls or walls of different thickness to match and have a smooth surface
- Making additional space to conceal fixtures or structural elements within a wall
- Preventing dampness

An additional type of channel used primarily for acoustical performance is *resilient channel.* Resilient furring channels are used over both metal and wood framing to provide a sound-absorbent spring mounting for gypsum board. *Figure 24* shows resilient channel used on the bottom of back-to-back CFS floor joists. The channels should be attached as specified by the manufacturer. They not only improve sound insulation, but also help isolate the gypsum board from structural movement, minimizing the possibility of cracking. Resilient furring channels may also be used for the application of gypsum boards over masonry and concrete walls. In wood frame construction, gypsum board can be screwed to resilient metal furring channels to provide a higher degree of sound control. Common types of steel furring channels include resilient, furring, and z-furring.

Figure 24 Resilient channel.
Source: Printed with permission from the American Iron and Steel Institute

Curves

One of the most beneficial features of steel framing is that it can be easily bent into curves. This allows for curved walls and dome-shaped roofs. Steel members may be purchased already bent into standard shapes, or they may be bent manually. Manual bending, however, is a tedious, time-consuming job that requires lots of measuring. Another option is to use a track bender tool. This device can be used on site to bend tracks quickly and accurately to the desired shape without measuring by simply setting a dial.

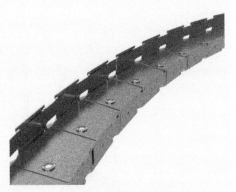

Source: Image provided by ClarkDietrich

1.2.3 Steel Floor Assemblies

Cold-formed steel floor assemblies typically use standard C-shape floor joists (*Figure 25*). Proprietary floor joists and pre-engineered steel floor trusses, *Figure 26*, may also be utilized. Equipment for installing steel floor assemblies include **rim track**, web stiffeners, **clip angles**, hold-down anchors, hangers, strapping, and fasteners. Like floor assemblies constructed with wood, steel floor systems can be installed directly onto a structure's foundation, connected by a wood sill, attached to a load-bearing wall, or connected to an I-beam.

Steel floor assemblies are similar to conventional framing in that they use either single or continuous span installation techniques. In single span floor construction, joists are supported at their installation points with no additional support beam in the middle. In the continuous span technique, joists have at least one additional support beam in the middle. Lapping, a type of continuous span installation in which joists are installed across a central girder and

Rim track: A horizontal structural member that is connected to the end of a floor joist.

Clip angles: L-shaped pieces of steel (normally having a 90-degree bend), typically used for connections.

connected to each other where their ends overlap, is sometimes used to increase stability. The type of span used, the structural loads that will be placed on the floor system, and the manufacturer's specifications determine the correct spacing of the floor joists. Steel floor assemblies can support all types of surfaces from wood to precast concrete to tile. In large commercial buildings, sheets of steel decking material are often used to create a support structure for the surface flooring. *Figure 27* illustrates the common techniques used in cold-formed steel floor framing.

Figure 25 C-shape floor joists.
Source: Don Wheeler

Figure 26 Proprietary floor truss.
Source: Don Wheeler

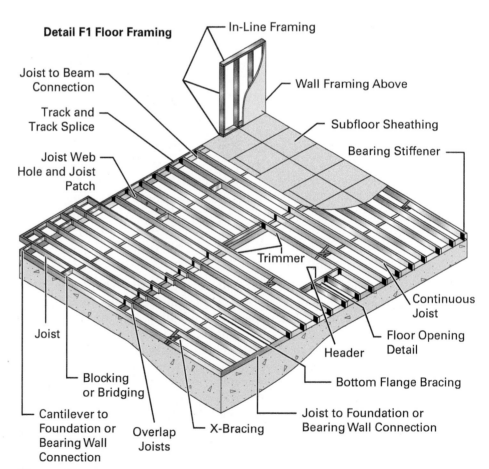

Figure 27 Steel floor framing techniques.

1.2.4 Steel Roof Assemblies

Recent years have seen a dramatic rise in the use of cold-formed steel framing members in roof assemblies. It allows for easy and standardized assembly, and the materials are durable and noncombustible. In addition to the standard C-shaped member, scores of proprietary shapes, fabrication methods and installation instructions are available from truss manufacturers nationwide. *Figure 28* shows an example of a pre-engineered steel roof system.

Figure 28 Steel roof system.
Source: Lev Kropotov/Shutterstock

The general features of a roofing system are very similar whether it is constructed of steel or wood. Rafters, ceiling joists, truss webbing, load-bearing studs, and various bracing and connecting hardware are used to either construct roof trusses on site or install prefabricated trusses. *Figure 29* illustrates the basic elements of a roof frame.

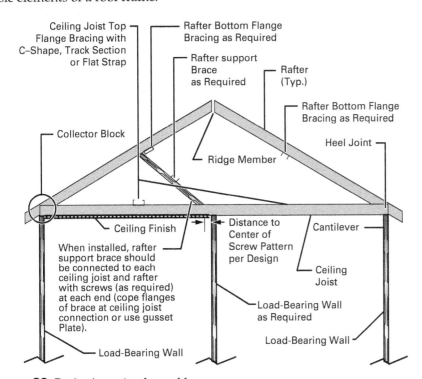

Figure 29 Basic elements of a roof frame.

Trusses are made from structural members that have been connected into triangular sections. Multiple trusses are used to make up the structural framework of a roof. Because all trusses must be constructed to meet strict standards, they are often pre-engineered rather than constructed in the field. When trusses must be built in the field, always measure and cut members accurately and connect them according to the specifications.

1.2.5 Ceiling Assemblies

Commercial ceiling systems often consist of a suspended metal grid in which ceiling tiles, usually acoustic ceiling panels, are placed. Detailed instructions for installing suspended acoustic ceiling systems can be found in NCCER Module 27209, *Suspended and Acoustical Ceilings*. When drywall is used as the surface material for a ceiling, a steel grid system is constructed as a framework to which the drywall will be attached. This framework may be suspended or directly mounted to the existing ceiling structure. The components of these systems are normally purchased as kits to be installed once piece at a time. *Figure 30* is an example of a pre-engineered drywall grid system. An alternative method is to use furring and cold-rolled channel to provide a rigid framework for suspended gypsum board and other ceiling assemblies, with the steel members suspended (*Figure 31*). The furring channel is perpendicular to the underside of the U-channel and is commonly clipped or wire-tied at appropriate intervals for attaching gypsum board with screws.

Figure 30 Pre-engineered ceiling grid system.
Source: Alex-White/Getty Images

Figure 31 U-channel ceiling framework.
Source: Don Wheeler

The furring channel can also be screwed to other structural steel members and can be attached directly to the bottoms of bar joists. In the latter case, the furring is installed perpendicular to the joists and wire-tied at appropriate intervals.

1.0.0 Section Review

1. A building must be able to support its dead load, live load, and other loads. Other loads are _____.
   a. the weight of the materials used to construct the building
   b. snow, rain, wind, and other forces of nature
   c. people and furniture inside the building once constructed
   d. the weight of the roof

2. Which method of framing structural walls involves constructing each wall on the jobsite, one stud at a time?
   a. Panelization
   b. Pre-engineered
   c. Stick building
   d. Fabrication

3. Which technique prevents cold-formed steel walls from twisting and buckling?
   a. Plenums
   b. C-shapes
   c. Racking
   d. Bracing

4. Nonstructural steel studs usually have a G40 galvanized coating weight, but structural studs have a galvanized coating weight of _____.
   a. G45 or higher
   b. G50
   c. G60 or higher
   d. G55

5. Cold-formed steel floor assemblies that are not proprietary or pre-engineered typically use _____.
   a. hat channel
   b. Z-shaped studs
   c. U-shaped studs
   d. C-shaped studs

6. True or False? The steel grid system used to mount drywall to a ceiling may be suspended or directly mounted.
   a. True
   b. False

2.0.0 Tools, Materials, and Construction Methods

Performance Tasks

1. Build a section of curtain wall from shop drawings to include a window opening with headers, jambs, and sill.

2. Build headers (back-to-back, box, and L-header).

3. Lay out and install a steel stud structural wall with openings to include bracing and blocking.

4. Lay out and install a steel stud nonstructural wall with openings to include bracing and blocking.

Objective

Describe tools, materials, construction methods used in steel framing.

a. Identify tools used for installing steel framing.

b. Identify fasteners used for installing steel framing.

c. Describe the purpose and types of steel framing connectors.

d. Describe steel materials, how to identify them, and basic construction methods for installing them.

The first section of this module examines the characteristics of cold-formed steel, its advantages as a framing material, and some of the many uses of cold-formed steel in construction. You may have noticed many fundamental similarities between traditional wood framing and steel framing. Joists, studs, beams, trusses, columns, and the like serve the same structural functions in both framing methods. Steel framing, however, requires a worker to master specialized tools and joining techniques. To do the work of a skilled steel framing professional, from installing walls to floor systems to roof trusses, you will need to be knowledgeable about the unique materials and installation methods of the trade.

2.1.0 Tools for Steel Framing Work

CAUTION

Always practice power tool safety. This includes wearing all of the appropriate personal protective equipment, ensuring power tools are properly grounded with three-pronged plugs, and inspecting tools and their cords for damage.

NOTE

A power screwdriver is not just a drill with a screwdriver bit. It normally has an adjustable depth control to prevent over-driving the screws. Many have a clutch mechanism that disengages when the screw has been driven to a preset depth. Some power screwdrivers are designed to perform specific fastening jobs, such as fastening drywall to walls and ceilings.

Working with cold-formed steel requires the use of a variety of specialized tools. The toughness of steel makes it a great building material, but cutting, drilling, and fastening can be challenging if you are not using the right tools. Steel framers must be very knowledgeable about the tools of their trade in order to make informed decisions about the following:

- Which tool is the most appropriate for a given task
- How to operate tools skillfully and effectively
- How to use tools safely

Some of the common tools that are used in steel framing component assembly include the following:

- *Screw gun* — This tool (*Figure 32*) is used to connect steel members together. A screw gun can also be used to attach sheathing material, such as plywood and gypsum board, to steel. However, a specialized drywall screw gun should be used when attaching gypsum board as it allows you to set the penetration depth to prevent driving too deep and damaging the drywall. A screw gun is also called a power screwdriver. It increases the speed and efficiency of construction work. The preferred screw gun for structural steel-to-steel connections should have an adjustable clutch and torque setting, with a speed range of 0 to 2,500 rpm. An impact screw gun can also be used for metal framing. When using an impact screw gun, make sure that the steel being secured is at least 33 mils in thickness and the screws are #10 in diameter or larger. A drywall screw gun (0 to 4,000 rpm) is recommended for attaching plywood or gypsum board to nonstructural steel less than 33 mils thick. Feed attachments are also available that automatically supply screws into the gun (*Figure 33*).

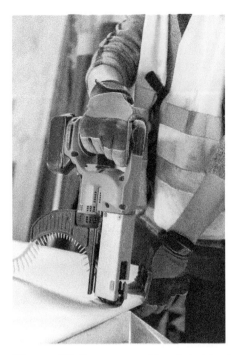

Figure 32 Cordless screw gun.
Source: Courtesy of Stanley Black & Decker, Inc.

Figure 33 Screw gun with collated feed.
Source: brizmaker/Getty Images

- *Powder-actuated fastener (PAF)* — The powder-actuated fastener tool, sometimes called a shot gun (*Figure 34*), is used to drive fasteners into concrete slabs, structural steel, and foundations. These tools get their power from a controlled explosion that occurs each time the trigger is pulled. Operators must be properly trained and have an operator's certificate before using this tool.

- *Hammer drill* — Hammer drills (*Figure 35*), as the name implies, apply a hammering action to the drill bit that is highly effective at drilling into masonry and concrete. This tool would be appropriate for attaching steel to brick, stone, concrete, or similar materials.

Figure 34 Powder-actuated fastener.
Source: Courtesy of Stanley Black & Decker, Inc.

Figure 35 Hammer drill.
Source: Courtesy of Stanley Black & Decker, Inc.

- *Impact driver* — Impact drivers, sometimes referred to as impact screw guns or high impact guns, are also used with cold-formed steel. However, they should only be used with steel that is 33 mils or thicker and screws that are at least #10 in diameter to avoid stripping out screws and weakening screw heads.

> ### WARNING!
>
> Powder-actuated tools can cause serious injuries if used improperly. Always wear a hard hat, heavy boots, and eye, ear, and face protection when using these tools. Treat a powder-actuated tool as if it were a loaded gun. Before you handle it, determine whether it is loaded with a powder charge and a fastener—either can be dangerous. Always point the tool away from yourself and others. Make sure that you know when the tool is in the locked and unlocked positions.
>
> Powder-actuated fasteners can be used only by personnel trained and certified in the use of the specific tool. You must carry your certification card with you when using the tool, and the card must be valid for the tool you are using.
>
> Be sure to select the proper charge for the job in accordance with the manufacturer's instructions.

- *Locking C-clamp* — Clamps are used to hold steel members securely together during fastening, which is especially important in steel framing. This is because as the screw penetrates the first layer of steel, it tends to pull that layer away from the layer to which it is being attached. This tendency is called *climbing the threads*. You may also hear it referred to as *screwjacking*. When it happens, it leaves a gap between the members that can decrease the strength of the joint. To prevent this, clamp the members together with the clamp tips positioned as close as possible to the fastening point. Use as many clamps as are needed to secure the layers. All clamps used in steel framing should have regular tips on the ends, without fixed or swivel pads. This allows all of the pressure to be placed on a small area instead of being dispersed across the pads.

Clamps come in a variety of sizes. C-clamps (*Figure 36*) are commonly used because their shape allows them to reach easily around cold-formed steel members.

Figure 36 Locking C-clamp.
Source: Dennis Axer/Alamy Images

2.1.1 Cutting Tools

Prefabrication of steel framing materials minimizes the amount of steel cutting that has to be done in the field. However, when field cutting must be performed, consider using the following tools and methods:

- *Circular saws* — Chop saws, heavy-duty circular saws that have abrasive blades, are commonly used for field cutting (*Figure 37*). Abrasive blades grind through metal instead of cutting it. While chop saws are highly effective for square cuts and for cutting bundled studs, they are very noisy and produce a rough-cut edge. In addition to chop saws, there are a number of other circular saws on the market that can also be used, with the right type of saw blade, to cut steel studs. Ferrous blades like carbide-tipped blades (*Figure 38*) and diamond-tip blades are used for cutting hard, iron-containing metals like steel. Ferrous blades provide a cleaner cut through steel than abrasive blades. A miter saw, which is similar to a chop saw but with the ability to make angled cuts, can also be used (with the appropriate saw blade) to cut cold-formed steel.

Figure 37 Cutting steel with a chop saw.
Source: Susan Law Cain/Shutterstock

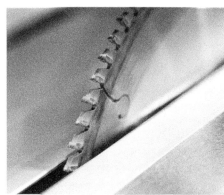

Figure 38 Saw blade with carbide teeth.
Source: Eugene Sergeev/Alamy Images

- *Dry-cut metal saws* — Another option for producing clean cuts in the field is to use a specialized, nonabrasive cutting tool. Dry-cut circular metal saws use blades made of carbide, titanium, or aluminum that cut through cold-formed steel and leave a smooth edge. The cutting action of this saw is usually slower than that of a saw with an abrasive blade, but the clean cut and reduction of hot, flying debris make it safer to use.

• *Swivel-head shears* — Swivel-head shears are available in both electric and battery-operated models that can cut many materials up to 68 mils thick, also known as 14-gauge. One mil is equal to $^1/_{1,000}$ inch. Swivel head shears are used to cut steel studs, runners, sheet metal, and cold-formed steel. They are portable and make smooth cuts with no abrasive edges. The cutting edge of the swivel-head shear rotates 180 degrees to make overhead and side work easier. Disadvantages include difficulty in cutting tight radii on C-shapes, and expensive replacement blades.

• *Aviation snips* — Aviation snips (*Figure 39*) are hand tools that can cut cold-formed steel up to 33 mils in thickness. They are useful for cutting and coping steel, snipping flanges, and for making small cuts without leaving abrasive edges. Aviation snips are color coded depending on the direction of the cut they are used to make.

Right (Clockwise) Cuts   Left (Counterclockwise) Cuts   Straight Cuts

Figure 39 Aviation snips.
Source: Ridge Tool Company

WARNING!

It is sometimes necessary for other trades to cut additional openings into steel framing. Field-cut holes must follow the requirements in AISI's S240, *North American Standard for Cold-Formed Steel Structural Framing*. If holes are not cut properly, the integrity of the structural steel members may be compromised.

Source: pioneer111/Getty Images

Did You Know?

Evolution of Power Tools

Power tools have come a long way since the first portable electric hand tools came on the scene around 1895. In 1961, Black & Decker was credited with the invention of the first cordless drill, which was powered by a nickel-cadmium battery. In 1978, Makita introduced a cordless drill with a removable nickel-cadmium battery pack. Another big advancement in portable power tools took place in 2005 when hand tools powered by the smaller but mightier (and more environmentally friendly) lithium-ion batteries made their appearance. While corded power tools have always had their advantages (continous power supply and higher power output), improvements in battery technology are quickly making battery-powered tools more powerful and convenient. Screw guns, impact drivers, hammer drills, swivel-head shears, and even saws can be powered with lithium-ion battery packs.

2.1.2 Laser Devices

Although conventional bubble levels have not disappeared from jobsites completely, the accuracy and efficiency of laser levels are making them the tool of choice. Laser levels (*Figure 40*) emit a beam of light from a device mounted on a base or tripod. Self-leveling laser levels automatically adjust the light beam so that it projects a level line onto a surface. Manual laser levels have to be adjusted by the user by adjusting thumb screws or bubble vials. Some laser levels also project plumb dots, which allow the user to transfer points from floor to ceiling with a high degree of accuracy. Rotary laser levels emit a continuous horizontal or vertical 360-degree beam all the way around an area, which can be used to rapidly measure large areas and take elevation readings.

As with the manual methods, the accuracy of a laser level is susceptible to human error. Take your time when setting up a laser level to ensure precision. Make sure the tripod or base of the level is on a flat surface and that it does not move during the course of your leveling task.

CAUTION

Always exercise caution when using any laser device. Never use a laser device that you are not trained to use. Some devices require a special operator's training and certification before use. Never shine the laser in anyone's eyes or toward a roadway. Even in daylight, a laser can temporarily blind a person. Always post warning signs in conspicuous locations when a laser is in use in the work area. Although most laser devices do not require the use of eye protection, always check the manufacturer's recommendations before use.

Figure 40 Laser level.
Source: Vitalliy/Getty Images

2.2.0   Steel Framing Fasteners

When selecting fasteners, steel framers must consider the loads to be transmitted through the connection, as well as the thickness, strength, and configuration of the materials to be joined. Fasteners include screws, nails, pins, clinches, and welds, as well as anchor bolts, rivets, powder-actuated fasteners, and expansion bolts. It is important to follow manufacturer recommendations for fastening systems during all phases of cold-formed steel construction projects.

Many factors affect the holding strength of fasteners. These factors include the characteristics of the building material to which the steel is being fastened (compressive strength of concrete for example), the fastener shank diameter, and how deep the fastener penetrates into the material. Fastener holding strength is also determined by the spacing between fasteners and the condition of the material. Steel framers use specialized fasteners to increase holding strength.

2.2.1 Screws

In steel framing, the screw is the workhorse of the fasteners. Screws are installed with screw guns, and they come in a number of different head styles to fit the wide range of structural and cosmetic requirements. Screws have three distinct thread thicknesses: coarse (threads spaced farthest apart), medium, and fine (threads spaced closest together). In steel framing, screws with coarse threads are generally used because they are best for cutting through steel. Thicker steel (97 mils for example) requires the use of screws with a fine thread because it increases the connection points between the screw and the material. The fine threads make the screw easier to drive and create a stronger hold.

Screws are usually finished with zinc or cadmium plating to withstand environmental impacts and resist corrosion. The size and the type of screw needed will depend on the thickness of the sheathing material and steel. Familiarity with the different types of screws that are available is important. If the wrong screw is used, it may not penetrate the steel, or it may break later, causing the joint to fail.

Steel framing is not typically predrilled for attachment screws. As a result, the screws used to secure steel framing must be able to make their own holes as they are placed. There are two main types of steel framing screws: self-drilling and self-piercing (*Figure 41*). Self-drilling screws do not have threads running the entire length of their shafts to the tip. Instead, they have a drill bit on the tip. Self-drilling screws are designed to cut through layers of steel and make a hole before any of the screw threads engage. Self-piercing screws have threads running the length of their shafts to the tip. The tip is a sharp point that can typically penetrate up to 33 mils of material with ease. Winged self-drilling screws (*Figure 42*) have extensions that bore a larger hole into the material being attached to steel and then break off once they contact steel. Self-piercing screws are commonly used to attach plywood and gypsum board to the thinner layers of steel.

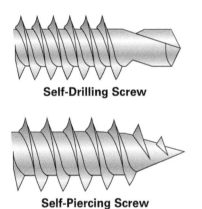

Self-Drilling Screw

Self-Piercing Screw

Figure 41 Self-drilling and self-piercing screws.

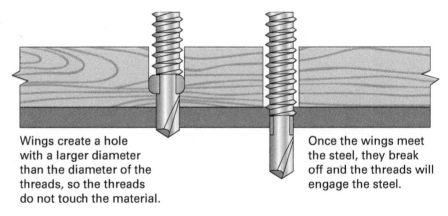

Wings create a hole with a larger diameter than the diameter of the threads, so the threads do not touch the material.

Once the wings meet the steel, they break off and the threads will engage the steel.

Figure 42 Winged self-drilling screws.

Code and standard requirements state that for all connections, screws must extend through steel a minimum of three exposed threads (*Figure 43*). For most steel-to-steel connections, $\frac{1}{2}$-inch or $\frac{3}{4}$-inch screws are acceptable. When applying plywood, gypsum board, or foam insulation to steel members, the proper screw length is found by adding the measured thickness of all materials and then adding $\frac{3}{8}$ inch to allow for the three exposed threads, as in the following equation:

Proper screw length = measured thickness of all materials + $\frac{3}{8}$ inch

Screws are available in many different head styles. The head is used to lock the screw into place and prevent it from sinking through the layers of material it is securing. The head style determines the type of bit tip used to place the screw. The proper selection of a head type profile ensures that the screw will work well with other building components.

Common self-drilling screws (*Figure 44*) include the following:

- *Gypsum board (bugle-head) screw* — Used to attach gypsum board to steel framing. The gypsum board screw should be used with a depth-setting nosepiece on the drill to avoid tearing the protective paper.

- *Flat-head screw* — Used for wood flooring and facings. The flat-head screw countersinks into the wood so that the top of the head is seated flush with the wood surface without causing the wood to splinter or split.

CAUTION

When securing screws, avoid drilling in so far that the screw head countersinks below the surface of the material being attached. This is known as *over-drilling*, and it will reduce the effectiveness of the connection and hurt the structural integrity of the sheathing.

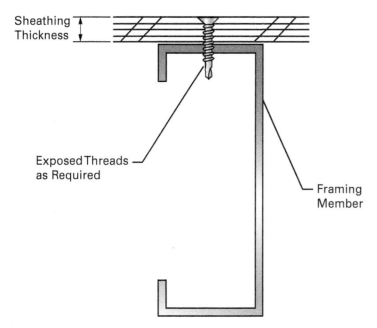

Figure 43 Proper screw attachment.

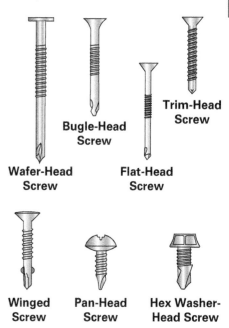

Figure 44 Typical self-drilling screws.

- *Wafer-head screw* — Has a larger head than the flat-head screw, but it is much thinner. The wafer-head screw seats almost flush with the surface in which it is placed, giving a clean, finished look. The large head gives it a greater surface area than other screws. For this reason, it is used for attaching to steel studs those materials that can be easily torn, such as felt or reflective foil.

- *Hex washer-head screw* — This screw is the favorite of cold-formed steel framers due to its tendency to stay put in the magnetic nosepiece of the screw gun. The head is hex-shaped and is also flared at the bottom, which serves as a washer. These screws are placed using a hex-head driver socket. The driver socket fits securely over the screw head, providing good stability while the screw is being driven into place. Hex washer-head screws should not be used beneath gypsum wall board or other sheathing that will be finished because they do not have low-profile heads.

- *Trim-head screw* — Can be used to attach baseboards and trim to steel studs, although they tend to split the wood and are not typically recommended. The small head penetrates the trim, leaving a tiny hole that can be filled with putty. It is recommended to use specially formulated trim nails and adhesives instead.

- *Winged screws* — Have wings that create a larger hole in the wood and then break off once the screw contacts the steel. This allows the threads of the screw to connect properly with the steel.

- *Pan-head screw* — Used to fasten studs to runner tracks. The pan-head screw is also used to connect steel bridging, strapping, or furring channels to studs or joists. The low-profile head of pan head screws make them a good choice for use under gypsum board that will be finished.

2.2.2 Pins

Pins resemble nails that have a spiraled or **knurled** shank (*Figure 45*), which increases holding power. Pins can be used for steel-to-steel connections or to attach sheathing material to steel members. Connections are made by driving the pin into the layers of material. Pins are commonly used to attach plywood, oriented strand board (OSB), and siding to steel framing. Some contractors use only pins, while others use screws around the edges and pins in the field of the board. Plywood or OSB sheathing material may be applied using pneumatic pins, as shown in the second example of *Figure 45*. To securely attach the sheathing, it must be held tightly against the steel before the pin is driven. This is because firing the pin does not pull the plywood against the steel as a screw does.

NOTE

Fasteners loaded under tension may require washers to prevent them from pulling through the thinner, nonstructural framing.

Knurled: A series of small ridges, dimples, or embossments used to provide a better gripping surface on metal.

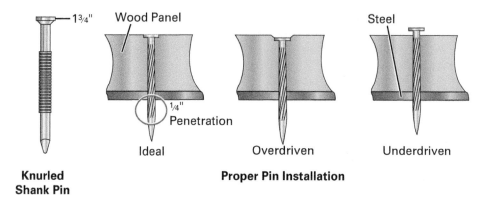

Figure 45 Pin types and pin installation.

Roof or floor sheathing is more easily installed with pins than wall sheathing because, on the roof or floor, the worker usually stands on the plywood to hold it tightly against the steel. Sheathing may also be attached to steel studs by first tacking the material to the steel with screws along the edges and then applying pins to the field. When attaching wood to steel, specifically formulated nails and adhesives should be used. Trim nails are designed to secure baseboards and other types of moulding without damaging the wood. Pneumatic and powder-actuated trim nailers are used with trim nails. Some manufacturers make trim nailers that can penetrate up to a double layer of 54 mil steel.

2.2.3 Press Joining and Clinch Joining

Press joining and clinch joining require no screws or pins, only a tool that presses two different sections of metal together, creating a metal-to-metal connection. Press joining tools (*Figure 46*) have a variety of applications including cold-formed steel framing and plumbing. Clinching tools are typically used with sheet metal and use a slightly different process. Manufacturers of clinching tools provide test reports that verify the strength of the press joining. Clinching systems work well on steel panels and are less permanent than welding. If studs are fastened incorrectly with a clinch connection, the connection must be drilled or cut out. Clinching connections are typically created in a shop setting and would rarely be done in the field.

Figure 46 Press joining tool.
Source: Ridge Tool Company

2.2.4 Bolts and Anchors

Bolts and anchors are also commonly used to fasten cold-formed steel framing to masonry, concrete, and other steel materials. Except for some proprietary anchors, pre-drilling of holes is necessary. Bolts require the installation of a washer and must meet or exceed the requirements of AISI S100. Expansion anchors are commonly used for connections to concrete or masonry and require information from the manufacturer to determine the capacity and spacing requirements. *Figure 47* illustrates how an anchor bolt would be used to secure a section of cold-formed steel floor frame to a foundation.

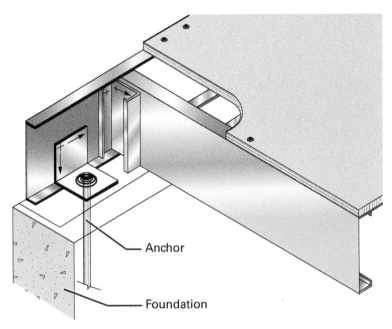

Anchor

Foundation

Figure 47 Anchor bolt securing floor to foundation.

2.3.0 Steel Framing Connectors

Steel framers use a wide variety of connectors for the various commercial, mid-rise, and residential components. Connectors are engineered to perform specific functions including bridging, bracing, deflecting, and drifting. Steel framing floor joists, roof trusses, and door and window assemblies also have specialized connectors.

Slip connectors (*Figure 48*) are devices that allow for the vertical movement of a structure to prevent loads from being transferred to steel framing components that are not engineered to withstand vertical stress, such as curtain walls. Taking a look at the leftmost example of *Figure 48*, you will notice that one side of the slip connector is fastened to prevent horizontal movement while the other side is fastened so that the attached stud can shift slightly in a vertical direction. Slip connectors are used when a structural system other than steel framing is needed to carry the loads of upper floors and the roof down to the foundation. Loads on the upper parts of a structure cause it to deflect, which in turn may induce loading in walls and wall structures that are not designed to carry these loads. Slip connectors are designed to allow for this movement without creating unmanageable stress on components.

Slip connectors are generally located at the top of a wall panel, where it meets the underside of a structural element, such as a floor slab or beam. Under gravity, seismic, or wind loads, this upper portion of the structure may deflect up or down. The connector is designed to allow this movement, restrain the wall system from out-of-plane movement, and prevent any additional axial loading on the stud.

Slip connections are also useful at locations where a wall system is continuous, bypassing intermediate floors. When this occurs, a slip connection extends from the

Figure 48 Slip connectors.
Source: Image provided by ClarkDietrich

side of the structure and supports the wall laterally. Connections are also installed at roof bypasses where the wall system extends past a roof structure to form a parapet or high wall. At such locations, the slip connection must permit movement of the structure either up, due to wind uplift, or down, due to gravity loads.

In multistory construction, movement may occur at the floor below a curtain wall system, causing the entire wall to move down. When this happens, the connector at the top of the wall (either a bypass or below structure) must have sufficient capacity to allow the wall system to move down without creating tension on wall components.

2.3.1 Other Typical Connections

Other than slip connections, the most typical connections used in stud curtain-wall design include the following:

- *Base connections* — These connections may be stud to track, track to concrete, or steel deck.
- *Head or sill-to-jamb connections* — Points where the head or sill is directly connected to the jamb.
- *Continuous angle connections* — A field-welded connection usually found at spandrel framing.
- *Clip angle connections* — Typically found at non-slipped bypass or spandrel framing connections; designed to carry both lateral and self-weight forces.
- *Outrigger clips* — Short lengths of angles designed as an axially-loaded strut to carry stud lateral reactions back to the structure.
- *Wind girt connections* — Usually located at taller spandrel framing cases when the use of diagonal braces is impractical. These connections are not usually required to carry any gravity loads.
- *Diagonal braces or kickers (as opposed to X-bracing for shear walls)* — These stud diagonals provide for a bottom spandrel stud reaction back to the structure.

- *Stud-to-stud connections* — May be either lapped or track-to-track. These connections occasionally require movement allowance, which can be accommodated by a slip-track or slip-pin type detail.

- *Knee wall base* — This is a moment connection at the bottom of a knee or stub-wall. It is usually either a freestanding parapet or a long segment of wall under a continuous or ribbon window condition.

- *Deep leg track* — Also known as a *deflection track*. The deep leg track allows for some degree of vertical deflection of the wall, without the wall sustaining damage due to lateral movement. In this type of connection, the tops of the studs are not secured to the upper track. When the upper part of the structure is under load, the track moves down until it rests on the tops of the studs. When this method is used, the studs must be properly stabilized to prevent twisting and bending.

2.4.0 Steel Framing Materials and Methods

Steel framers must be familiar with a wide variety of materials including steel studs, runners (tracks), various types of steel channel, angles, joists, and steel roof trusses. Framing systems come with accessories such as clips, web stiffeners, resilient channels, fastening devices, and anchors required for complete and proper installation of members. Many manufacturers offer specialized products that enable builders to shorten construction times for complicated curves, arches, and other unique architectural features. The vertical and horizontal framing members are the structural load-carrying features for a large number of low- and high-rise structures.

As previously noted, the benefits of using steel framing materials include cost-efficiency, faster assembly and installation, and design flexibility. Cold-formed steel (CFS) framing materials are also noncombustible, lightweight, corrosion-resistant, and fabricated to fit together easily. Manufacturing tolerances for cold-formed steel members are governed by AISI standards. These standards govern length, web depth, flare, crown, bow, and twist. CFS is also coated against corrosion according to AISI standards. Hot-dipped zinc galvanizing is the most effective coating method. Depending on the thickness of zinc applied to the steel and the environment in which the steel is placed, zinc coatings can protect the steel for hundreds of years.

There are a variety of load-bearing and nonbearing systems. *Figure 49* shows the basic components of a steel stud system. To meet custom material

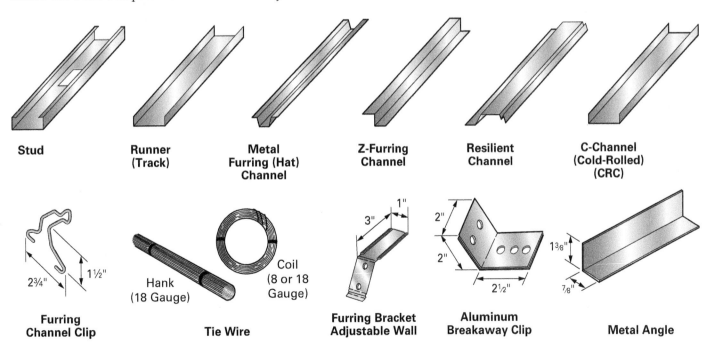

Figure 49 Components of a steel stud system.

requirements, studs, track, and joist material can be cut within $\frac{1}{8}$ inch of specifications. Length is restricted only by the mode of physical transportation—typically 50 feet for containers and flatbed trucks. Custom ordering allows for less field cutting and less labor and waste on the jobsite.

Steel framing members that are not marked with the required identification should not be used. The identification may be etched, stamped, inkjetted, or labeled every 96 inches along the member (*Figure 50*). Missing identification can result in project delays until materials are identified in accordance with appropriate standards.

Figure 50 Steel product label.
Source: Don Wheeler

Going Green

Recycled Steel

Modern steel production relies on two technologies: the basic oxygen furnace (BOF) and the electric arc furnace (EAF). The BOF process uses 25 to 35 percent old steel to make new steel. It produces sheet for products in which the main requirement is drawability; that is, products like automotive fenders, steel-framing members, refrigerator encasements, and packaging. The EAF process uses 95 to 100 percent old steel to make new steel. It is primarily used to manufacture products in which strength is critical, such as structural beams, steel plates, and reinforcement bars. No matter which process is used, the resulting steel product has a minimum of 25 percent recycled content. Industry-wide, the recycled content of steel averages 67 percent.

During the production process, molten steel is poured into an ingot mold or a continuous caster, where it solidifies into large rectangular shapes known as slabs. The slabs are then passed through a machine with a series of rolls that reduce the steel to thin sheets with the desired thickness, strength, and physical properties. The sheets are sent through a hot-dipped galvanizing process and are then rolled into coils that can weigh more than 13 tons.

2.4.1 Framing Material Identification

Steel framing members are selected for different applications depending on their properties and load-carrying abilities. The standards and codes that govern the selection of framing materials are particular and must be followed exactly to ensure structural integrity. Manufacturers of various steel framing products have published widely varying codes for classifying steel framing materials. To eliminate this confusion, the Steel Stud Manufacturers Association (SSMA) and the American Iron and Steel Institute (AISI) developed the Universal Designator System. The designator consists of four required sections of code.

The first section, from left to right, is the manufacturer's ID. The next section is the thickness of the member in mils. The thickness is followed by the coating identifier. The last part of the code is the yield strength.

NOTE

You may have noticed that there are five sections of code on the stud in *Figure 50*. The additional section not previously mentioned (800S162) is part of the AISI standard nomenclature for steel framing members. The number 800 designates an 8-inch member depth (expressed in $\frac{1}{100}$ inch), S represents the type of steel member (lipped channel, which could be a stud or joist), and 162 indicates a $1\frac{5}{8}$-inch flange. This information is often included by some manufacturers for clarity.

Using *Figure 50* as an example, we can tell that this stud was manufactured by MBA, has a thickness of 43 mils, meets the minimum coating requirements (CP60), and has a yield strength of 33 KSI.

STUFL is a useful acronym in the steel framing industry to help remember the following five basic shapes of steel framing members:

S = Stud or joist framing member with **lips**
T = Track section
U = U-channel or stud framing section without lips (also known as edge stiffeners)
F = Furring channels
L = L-header or angle

Figure 51 illustrates the five different STUFL shapes and includes web depth and flange width, which are a required part of the AISI Universal Designator Code. The dimension labeled D represents the web depth (sometimes referred to as *member depth*). Dimension B is a measurement of the flange width. As you can see in *Figure 52*, manufacturers typically print additional information about the product on the bundle label, including the flange width (162 in this case) and the web depth (600). Both flange width and web depth are expressed in hundredths of an inch.

Lips: That part of a C-shape framing member that extends from the flange as a stiffening element that extends perpendicular to the flange; also called *edge stiffener*.

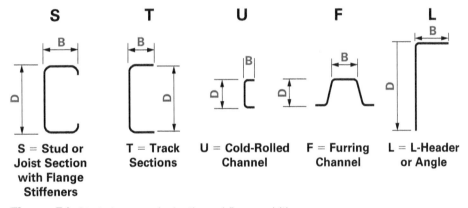

Figure 51 Stud shape, web depth, and flange width.

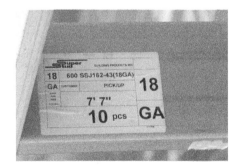

Figure 52 Manufacturer's product label.
Source: Don Wheeler

Base steel thickness was traditionally referenced in terms of gauge numbers, but that has given way to designating thickness in terms of mils. One mil is equal to $\frac{1}{1,000}$ of an inch. *Table 2* lists equivalent base steel thicknesses in mils and gauges as prescribed by the American Iron and Steel Institute standards S220 and S240, which are referenced in both commercial and residential building codes. Another dimensional relationship that steel framers should be familiar with is the dimension of the stiffening lip by flange width and material thickness as listed in *Table 3*.

Equivalent (EQ) Studs

Around 2005, manufacturers started developing more efficient and thinner steel products for nonstructural applications. To produce a stronger, lighter product, steel makers used higher yield-strength steels with more bends and folds in patented configurations. This new and improved steel was equivalent to, and in some cases better than, traditional steel in terms of strength and performance. Equivalent, or EQ, studs have become the new standard for interior framing.

CAUTION

EQ Studs are identified by *mils*, not *gauge*. If you encounter architectural plans or specifications that mention gauge instead of mils, be sure to clarify what the actual intended thickness is in mils. Manufacturers of EQ studs provide detailed information and support for their products. Use these resources to make sure EQ studs are approved before using them on a project.

TABLE 2 Minimum Base Steel Thickness of Cold-Formed Steel Members

Designation (Thickness in Mils)	Minimum Base Steel Thickness Inches (mm)[1]	Old Reference Gauge Number[2]
18	0.0179 (0.455)	25
27	0.0269 (0.683)	22
30	0.0296 (0.752)	20 – Drywall[3]
33	0.0329 (0.836)	20 – Structural[3]
43	0.0428 (1.09)	18
54	0.0538 (1.37)	16
68	0.0677 (1.72)	14
97	0.0966 (2.45)	12
118	0.1180 (3.00)	10

[1]Design thickness shall be the minimum base steel thickness divided by 0.95.
[2]Gauge thickness is an obsolete method of specifying sheet and strip thickness. Gauge numbers are only a rough approximation of steel thickness and shall not be used to order, design, or specify any sheet or strip steel product.
[3]Historically, 20-gauge material has been furnished in two different thicknesses for structural and drywall (nonstructural) applications.

TABLE 3 Stiffening Lips on C-Sections

Material Thickness	Flange Width	Stiffening Lip
Up to 33 mils (or 0.033")	$1\frac{1}{4}$"	$3/16$"
All thicknesses	$1\frac{3}{8}$"	$3/8$"
All thicknesses	$1\frac{5}{8}$"	$1/2$"
All thicknesses	2"	$5/8$"
All thicknesses	$2\frac{1}{2}$"	$5/8$"

2.4.2 Basic Construction Methods for Steel Framing

Steel framers must be well versed in the fundamental requirements and guidelines that apply to cold-formed steel construction. Common techniques and structural features that you will encounter as a steel framer include the following:

- In-line framing
- Web stiffeners
- Web holes and patches
- Built-up shapes
- Bridging, bracing, and blocking
- Insulation cavities

In addition to the essential elements, some special considerations and tricks of the steel framing trade can save you time and prevent costly mistakes.

In-Line Framing

In-line framing, or direct alignment, is the preferred and most common framing method for providing a direct load path for transfer from studs to joists, through the framing system, and to the ground. In this method, cold-formed steel framing members are aligned vertically so that the centerline of the joist web is within $3/4$ inch of the centerline of the structural stud member below, or the centerline of the stud web is within $3/4$ inch of the centerline of the web joist below (*Figure 53*).

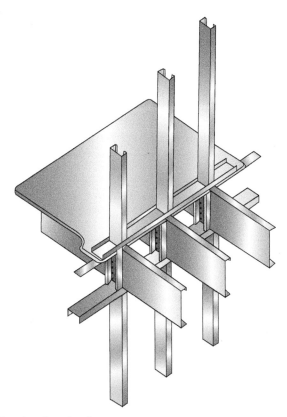

Figure 53 In-line framing detail.

Web Stiffeners

Web stiffeners, or bearing stiffeners (*Figure 54*), in the shape of C-shaped members or track members, are used to prevent joists from crippling at the point where the load transfers from a stud into the floor joist under structural walls. The thickness of a stiffener is, at a minimum, the same thickness as the floor joist, and the length of the stiffener is the depth of the joist minus $\frac{3}{8}$ inch.

$$\text{Length of web stiffener} = \text{Depth of the joist} - \tfrac{3}{8} \text{ inch}$$

Stiffeners are installed across the joist depth of the web and can be located on either side of the web. The stiffener is fastened to the web with either three or four No. 10 screws. Three fasteners are used when the screws are installed in a single row, and four are used when screws are installed with one fastener in each corner.

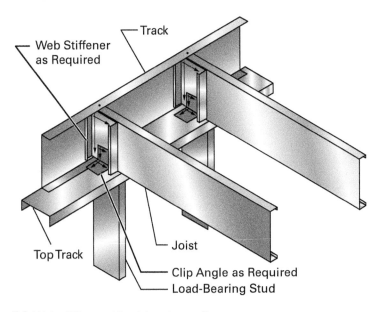

Figure 54 Web stiffener at load-bearing wall.

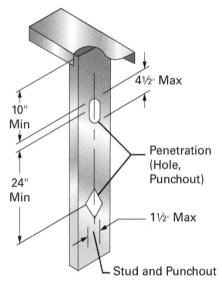

Figure 55 Standard stud punchouts (web holes).

Web Holes and Patches

Depending upon the application, framing members may be designed and manufactured with standard punchouts—web holes made during manufacturing (*Figure 55*). Holes in webs of studs, joists, and tracks must conform to an approved design. Standard web hole sizes are typically 1½ inches by 4 inches and are located on the centerline of the web of the member at 24 inches. Proprietary products are also produced with much larger holes that have been engineered and approved to ensure that the member's strength is not compromised.

Built-Up Shapes

The open, C-shaped cross-section commonly used in cold-formed steel framing provides a very low resistance to twisting. Built-up shapes, such as a nested stud and track assembly, provide the increased stiffness of a closed section as well as flat surfaces for attaching finish materials. Built-up shapes are commonly used for door and window jambs, headers, beams, and posts (*Figure 56*). The strength of built-up shapes is determined by the properties of the individual members, as well as the method of fastening.

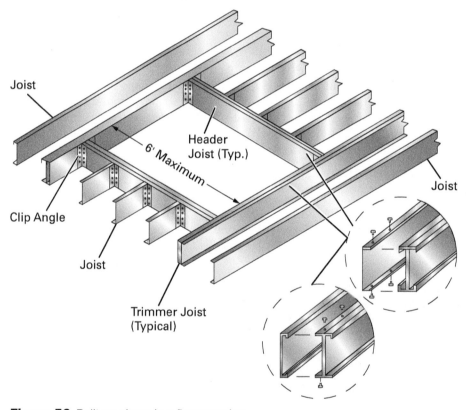

Figure 56 Built-up shape in a floor opening.

Bridging, Bracing, and Blocking

The strength of individual framing members is a function of the bracing provided. This bracing restrains the member from moving laterally or twisting, and it may be in the form of any one or a combination of the following:

- Cold-rolled channels placed through the web punchouts
- Steel strapping attached to the flanges with periodic solid **blocking** (*Figure 57*) or X-bracing
- Wood sheathing attached to the flanges

Blocking: C-shaped track, break shape, or flat strap material attached to structural members, flat strap, or sheathing panels to transfer shear forces.

Figure 57 Blocking.
Source: Image provided by ClarkDietrich

Overall, system bracing anchors the steel members and provides stability to the entire structure.

Insulation Cavities

Products used in construction may require some additional insulation to meet energy codes. The Thermal Design Guide for Exterior Walls, produced by the American Iron and Steel Institute, provides designers and contractors with guidance on thermal design of buildings that use cold-formed steel framing members. The thermal performance of a steel-framed structure may also be improved by the batt and other insulating materials within the wall cavity. Some thermal regions require insulation foam board on the exterior of the frame. Designers should consider the effects of moisture when assessing the application of cavity and continuous insulation.

Pressure-Treated Wood

Be careful when using pressure-treated wood products with cold-formed steel framing, as accelerated corrosion may result. It is preferable not to use pressure-treated wood with steel framing, but if you do, specify a less corrosive pressure treatment, such as sodium borate. Always separate the steel framing material from the pressure-treated wood with a non-absorbing, closed-cell sill seal.

Custom Jigs

The use of identical components, hardware, and accessories allows buildings to be constructed quickly and with little waste. For example, most doors in a building are identical, which saves time during construction. Each door opening will be the same size, each door jamb will be constructed in the same way, and all doors will use the same types of hardware.

When the same task is performed over and over, it is often helpful to construct a jig to help do the job. Examples of jigs include the following: a wooden frame that holds a steel stud upright while a permanent connection is being made; a template that is used to accurately drill holes in a steel member; and a frame that is used to square connections as the frame is being assembled. Jigs make doing the job faster and easier.

2.4.3 Assembling a Steel Frame

Steel framers not only install the structural framework, but they also often have a role in assembling these systems. Although pre-assembled floor systems, roof trusses, ceiling assemblies, structural walls, nonstructural walls, and a variety of other framing elements can be purchased from a manufacturer, some projects will require at least some portion of the assembly to be done on the jobsite. When an assembly is constructed in place, one piece at a time, it is known as *stick framing*. Alternatively, a component may be pre-assembled and brought to the installation location in a method referred to as *panelization*. Many projects that you work on as a steel framer will utilize a combination of panelization and stick framing. An overview of these two methods follows.

Stick Framing

Stick framing, as the name suggests, is a method of constructing the structure of a building stud by stud. In other words, framers are constructing the sections of the frame in place as opposed to using pre-assembled sections or panels. To frame walls in place, cut the track for the full length of the wall. Mark the layout using the same method described in Section 1.2.1 of this module. Once the layout has been marked, anchor the bottom track in place, securing the studs in each end of the track. Position the top track at the ends and with intermediate studs. Use a level to position the remaining studs for the wall. Install headers, X-bracing, or plywood with the wall standing.

Figure 58 provides a basic wall framing design. This type of wall may either be built in place or pre-assembled and then raised into position and secured to the building foundation. The spacing between the studs depends on the design of the building, the loads that will be placed on the structure, and the thickness of the steel being used. Typical stud spacing is 16 or 24 inches on center. Studs are bridged or strapped whenever there is a possibility that they will twist.

Depending on the design of the wall, it may need additional stabilizing. Shear walls, which need to withstand forces from wind and earthquakes, may need diagonal bracing. Other walls, such as interior nonbearing walls, may not require additional bracing; the sheathing attached to the studs may provide enough stability.

The possibility of vertical and horizontal deflection often figures into decisions about how the upper track will be secured to the wall studs. Sometimes the studs are screwed to the upper track, but this method does not allow for movement. If the roof or floor above deflects down too much, the screw and top track can transmit enough axial load to the studs to cause buckling and cracking of the wall below. To prevent joint failure, a method that allows for movement is used. One such technique is the use of slotted tracks (*Figure 59*). The screw holes in the sides of the track are vertical slots. The stud is attached to the track using a screw, which can move vertically in the slot. The studs are cut short—approximately $^3/_4$"—to allow that much downward movement of the structure above. The screws are typically installed in the center of the slot, to allow for downward movement of the lower floor or upward movement of the level above. When the stud is attached to the track, it can move up and down within the slot as the floor or roof above is placed under load.

Another method for controlling deflection is to use deep leg tracks. In this design, the sides of the track are deeper than those of a standard track. The deep leg tracks are placed over the tops of the studs, but the studs are not secured to the upper track. As a result, when the upper part of the structure is under load, the track moves down and does not transfer any vertical load to the studs below. When this method is used, the studs must be properly stabilized to prevent twisting and bending.

Movement of steel framing components can also be engineered using slip tracks. Although this configuration is difficult to build and is rarely used, it does still appear in some architectural details. In this method, the tops of the studs are attached to the upper track. The upper track is then nested in a second track, with no connection between the two tracks. The inner track is attached to the studs, which helps to stabilize the wall. The inner track floats inside the outer track as a vertical load is applied.

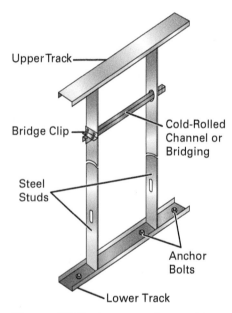

Upper Track

Bridge Clip

Cold-Rolled
Channel or
Bridging

Steel
Studs

Anchor
Bolts

Lower Track

Figure 58 Basic steel wall assembly.

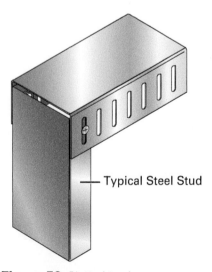

Typical Steel Stud

Figure 59 Slotted track.

Another process uses a variety of clips, called slip connectors, to connect members together. The clips are formed into a 90-degree angle, and slotted connector holes allow the connectors to move freely as a load is applied to the member. The different designs of the clips allow for horizontal drift or vertical deflection, or both. The clips may be used in a track, or they may be attached directly to stud and floor members.

Panel Framing

In panelized construction, walls are typically built on a framing table that allows the wall to be built horizontally at a comfortable working height. This is not a true "table" in a conventional sense but consists of a fixed bottom rail and an adjustable top rail. This adjustable top rail can be moved to accommodate different wall heights. It can also be used to compress the wall once studs are in place but before track connections are made, to help fully seat studs in the top and bottom tracks.

Many panel operations use welding rather than screw connections, which allows the stud-to-track connections to be made from the top side of the panel, without having to flip the panel over or have framers apply screws from the underside. Some panel tables have stops set up for standard 16" or 24" on center spacing to facilitate stud layout. Highly sophisticated tables will have laser layouts, which can project the location of each piece of the wall from a laser mounted above the table.

In a typical installation sequence, top and bottom tracks are set against the top and bottom rails of the table and held in place with clamps or magnets. Studs are installed between the top and bottom tracks. Sometimes built-up jamb studs and headers are pre-built at another location in the panel plant, so they can be dropped in place by the panel builders without slowing down the process to build these specialty elements. For exterior walls, the jambs and box headers are usually pre-filled with insulation when they are built, because this area will not be accessible once installed in the panel. Once everything is in place, for load-bearing panels the table is compressed (the top rail is moved towards the bottom rail) to help fully seat studs in the tracks. Once this is done, the panel is double-checked for squareness and stud location, and then the studs and tracks are welded or screwed together.

Some panel systems come with a machine that will pre-cut members to length, as well as provide dimples and holes at each screw location. This way, wall panel elements may be snapped in place without the use of the compression rails. Panel tables for this type of system consist of support beams for the studs and tracks, so panels can be built at waist height.

When the frame is raised and placed, it is attached to the building foundation by means of the lower track. Anchor bolts (*Figure 60*) are frequently set

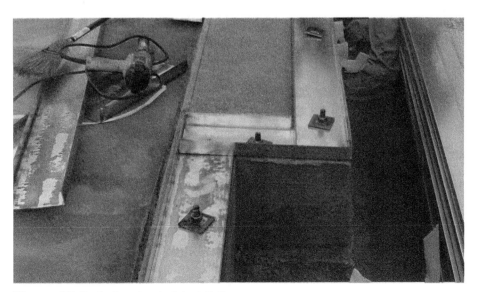

Figure 60 Typical anchor bolt installation.
Source: Don Wheeler

into the foundation before the concrete is poured. When no anchor bolts are present, they may be drilled in to create a connection point for steel framing. Anchor bolts allow the lower track of the steel framing to be bolted or welded to the foundation. Powder-actuated fasteners may also be used to attach the lower track to the foundation.

2.0.0 Section Review

1. A screw gun for making structural steel-to-steel connections should have an adjustable clutch and torque setting with a speed range of no more than _____.
   a. 1,500 rpm
   b. 2,500 rpm
   c. 3,500 rpm
   d. 4,000 rpm

2. Which type of laser level automatically adjusts the light beam so that it projects a level line onto a surface?
   a. Plum bob
   b. Self-leveling
   c. Manual
   d. Bubble

3. Which type of screw is favored by cold-formed steel framers due to its tendency to stay in place in a screw gun while being driven?
   a. Hex washer-head
   b. Pan-head
   c. Wafer-head
   d. Flat-head

4. Which type of connectors allow for the vertical movement of a structure without creating unmanageable stress on components?
   a. Clip angle
   b. Wind girt
   c. Kicker
   d. Slip

5. Which of the following provide increased stiffness as well as flat surfaces for attaching materials and are commonly used for door and window jambs, headers, beams, and posts?
   a. Built-up shapes
   b. C-shapes
   c. Slip connectors
   d. X-braces

6. Which technique, also known as direct alignment, is the preferred method for providing a direct load path through the framing system to the ground?
   a. Stick framing
   b. In-line framing
   c. Panelization
   d. Bracing

1. How many inches apart from the center of one stud to the next are structural studs usually placed?
    a. 18
    b. 12
    c. 24
    d. 16

2. A steel framing component that has a web, two flanges, and two lips is called a _____.
    a. cripple stud
    b. C-shape
    c. web stiffener
    d. header

3. What type of wall is designed to resist lateral forces such as those caused by earthquakes or wind?
    a. Curtain
    b. Panel
    c. Shear
    d. Curved

4. The two main methods for bracing a shear wall are X-bracing and _____.
    a. structural sheathing
    b. blocking
    c. cross bracing
    d. balloon framing

5. The thickness of nonstructural framing members ranges from _____ mils.
    a. 12 to 18
    b. 15 to 33
    c. 30 to 33
    d. 33 to 47

6. Which of the following consist of connected triangular sections of steel structural members and are often pre-engineered?
    a. Roof trusses
    b. Ceiling assemblies
    c. Floor trusses
    d. Shear walls

7. Which of the following is used to provide a rigid framework for drywall and other ceiling assemblies?
    a. Slotted track
    b. X-bracing
    c. Cold-rolled channel and furring channel
    d. Cold-rolled channel and diagonal strapping

8. What type of screw gun allows you to set the penetration depth to avoid driving a screw too deep?
    a. Impact screw gun
    b. Standard screw gun
    c. Drywall screw gun
    d. Locking screw gun

Figure RQ01
Source: Dennis Axer/Alamy Images

9. The tool shown in *Figure RQ01* is a(n) _____.
   a. channel stud shear
   b. locking C-clamp
   c. aviation snip
   d. stud crimper

10. Which type of saw blades cut steel and leave clean, smooth edges?
    a. Nonabrasive
    b. Noncorrosive
    c. Abrasive
    d. Standard

11. A cutting tool that can cut cold-formed steel up to 33 mils thick and leave no abrasive edges is a(n) _____.
    a. chop saw
    b. swivel-head shear
    c. C-clamp
    d. aviation snip

12. One mil is equivalent to _____.
    a. $\frac{1}{10}$ of an inch
    b. $\frac{1}{100}$ of an inch
    c. $\frac{1}{1,000}$ of an inch
    d. $\frac{1}{1,000}$ of a foot

13. When selecting a screw to attach steel framing parts, the screw length should be the thickness of all the material plus _____ inch.
    a. $\frac{1}{8}$
    b. $\frac{1}{4}$
    c. $\frac{3}{8}$
    d. $\frac{1}{2}$

14. What is the purpose of spiraled and knurled shanks on pins?
    a. To make them easier to drive
    b. To increase their holding power
    c. To make them faster to drive
    d. To decrease the friction between the components being attached

15. When using bolts to fasten steel framing to concrete, you must install _____.
    a. a washer
    b. a hex washer-head screw
    c. caulking
    d. pneumatic pins

16. What was created to eliminate the confusion caused by the wide variety of codes for classifying steel framing materials that have been published by manufacturers of steel framing products?
    a. American Iron and Steel Institute
    b. Universal Product Code
    c. Universal Designator System
    d. Steel Stud Manufacturers Association

17. According to the STUFL acronym, the five basic types of steel framing members are studs, track sections, U-channel or edge stiffeners, furring channels, and _____.
    a. furring channel
    b. lipless studs
    c. C-channel
    d. L-headers or angles

18. The part of a C-shape framing member that extends from the edge of a flange as a stiffening element is called a _____.
    a. lateral
    b. lip
    c. web
    d. knurl

19. The maximum centerline to centerline tolerance for in-line framing of cold-formed steel is _____ inch.
    a. $\frac{1}{8}$
    b. $\frac{1}{4}$
    c. $\frac{1}{2}$
    d. $\frac{3}{4}$

20. Which method makes use of pre-assembled sections or panels which are moved from one location to another to be installed?
    a. Engineering
    b. Structuring
    c. Panelization
    d. Building-up

Answers to odd-numbered review questions are found in the Module Review Answer Key found at the end of this book.

Answers to Section Review Questions

Answer	Section	Objective
Section One		
1. b	1.1.0	1a
2. c	1.2.1	1b
3. d	1.2.1	1b
4. c	1.2.2; Table 1	1b
5. d	1.2.3	1b
6. a	1.2.5	1b
Section Two		
1. b	2.1.0	2a
2. b	2.1.2	2a
3. a	2.2.1	2b
4. d	2.3.0	2c
5. a	2.4.2	2d
6. b	2.4.2	2d

MODULE 45203

Suspended and Acoustical Ceilings

Source: Cynthia Lee/Alamy

Objectives

Successful completion of this module prepares you to do the following:

1. Identify the components necessary to properly install a suspended ceiling system.
   a. Identify the suspension systems and hardware necessary to properly install a suspended ceiling system.
   b. Identify the system components necessary to properly frame a suspended ceiling system.
   c. Identify the safe material handling and storage procedures required when installing a suspended ceiling system.
2. Interpret a reflected ceiling plan.
   a. Interpret the layout information.
   b. Interpret the mechanical, electrical, and plumbing (MEP) locations.
3. Describe the key considerations, methods, and best practices relating to ceiling installation.
   a. Explain seismic considerations for ceilings.
   b. Identify the layout and takeoff procedures to install a suspended ceiling system.
   c. Identify the tools and equipment to lay out and install a suspended ceiling system.
   d. Identify the installation methods and procedures for a suspended ceiling system.

Performance Tasks

Under supervision, you should be able to do the following:

1. Estimate the quantities of materials needed to install a lay-in suspended ceiling system in a typical room from an instructor-supplied drawing.
2. Establish a level line at ceiling level, such as is required when installing the wall angle for a suspended ceiling.
3. Lay out and install a lay-in suspended ceiling system according to an instructor-supplied drawing.

Overview

Suspended ceilings are found in most commercial buildings. Unlike fixed ceilings, they provide easy access to wiring, cabling, and air conditioning equipment located in the area between the ceiling and the overhead deck. The ceiling tiles suppress sound transmission. Suspended ceilings are built by first installing a grid suspended from the overhead deck, and then installing ceiling tiles in the grid. Proper installation of these ceilings is a skill that takes training and practice.

1.0.0 Suspended Ceiling System Components

Performance Tasks

There are no Performance Tasks in this section.

Objective

Identify the components necessary to properly install a suspended ceiling system.

a. Identify the suspension systems and hardware necessary to properly install a suspended ceiling system.

b. Identify the system components necessary to properly frame a suspended ceiling system.

c. Identify the safe material handling and storage procedures required when installing a suspended ceiling system.

Ceiling panels: Acoustical ceiling boards that are suspended by a concealed grid mounting system. The edges are often kerfed and cut back.

Ceiling tiles: Any lay-in acoustical boards designed for use with exposed grid mounting systems. Ceiling tiles normally do not have finished edges or precise dimensional tolerances because the exposed grid mounting system provides the trim-out.

Acoustical materials: Types of ceiling panel, plaster, and other materials that have high absorption characteristics for sound waves.

Suspended ceilings are widely used in commercial construction and to some extent in residential construction. Modern suspended ceilings serve many purposes. They are designed to help keep outside noise from entering the room and to reduce noise levels occurring within the room itself. In some cases, ceilings are integrated with the electrical and HVAC (heating, ventilating, and air conditioning) functions to provide correct lighting and temperature control. Use of attractive **ceiling panels** or **ceiling tiles** helps give a warm, relaxed feeling to a room. Complete ceiling systems offer a wide variety of options, both functional and visual. The type of **acoustical materials**, the plans for the acoustical ceiling, and the method of installing the ceiling depend on the intended use of the room.

Sounds travel through the air in a room as a series of pressure waves. These sound pressure waves travel outward in all directions. When sound waves strike a wall or ceiling, some of the sound-wave energy is absorbed and some is reflected in wave patterns moving in the opposite direction (*Figure 1*). The result

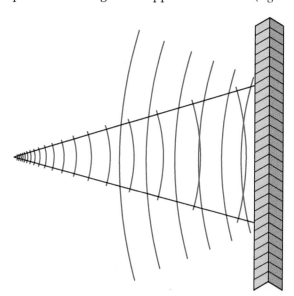

Figure 1 Sound-wave reflection.

is that you will hear both the original sound and its reflected image. The sound is also transmitted through the air in the wall or ceiling cavity to the opposite surface, causing the surface to vibrate and transmit the sound to any adjoining room(s).

Sound waves have a frequency. The frequency, or pitch, measured in hertz (Hz), is the number of vibrations or cycles that occur in the wave in one second. The greater the number of cycles per second (Hz), the higher the frequency and the higher the pitch.

The intensity of sound refers to its degree of loudness or softness. An A-weighted decibel (dBA) is a unit of measure used for establishing and comparing the intensity of sound sources. It is used to express the value of all sounds in a range from 0 dBA to 140 dBA and higher.

Table 1 shows some typical examples of noise situations and their relative decibel (dB) levels. Note that any sounds greater than 120 dBA can produce a physical sensation. Sounds above 130 dBA can cause pain and/or deafness. Continued exposure to sound levels above 85 dBA can cause hearing loss over time.

Some terms you may encounter and should understand when selecting or working with acoustics and acoustical ceiling materials include the following:

- *Reflection* — The bouncing back of sound waves after hitting some obstacle or surface such as a ceiling or wall.

- *Reverberation* — The prolonging of a sound through multiple reflections of that sound as it travels back and forth across a room. These multiple reflections of sound occur so fast that they are usually not heard as distinct repetitions of that sound. However, they can cause a higher noise level than the original sound source.

- *Noise reduction coefficient (NRC)* — Used by manufacturers to compare the noise absorbencies of acoustical products. The higher the number, the better the absorbency. The NRC measures the average percentage of noise a material absorbs at four selected frequencies.

- *Sound transmission loss* — The amount of sound lost as a noise travels through a material. Acoustical ceiling assemblies are rated in terms of sound

Frequency: Cycles per unit of time, usually expressed in hertz (Hz).

Hertz (Hz): A unit of frequency equal to one cycle per second.

A-weighted decibel (dBA): A single number measurement based on the decibel but weighted to approximate the response of the human ear with respect to frequencies.

Decibel (dB): An expression of the relative loudness of sounds in air as perceived by the human ear.

Acoustics: A science involving the production, transmission, reception, and effects of sound. In a room or other location, it refers to those characteristics that control reflections of sound waves and thus the sound reception in the area.

TABLE 1 Sound Levels of Some Common Noises

Sound Level (dBA)	Intensity Level	Outdoor Environment	Indoor Environment
140	Deafening	Jet aircraft, artillery fire	Gunshot
130	Threshold of pain	—	Loud rock band
120	Threshold of feeling	Elevated train	Portable stereo headset on high setting
110	Extremely loud	Overhead jet aircraft at 1,000'	Loud nightclub
100	Very loud	Chainsaw, motorcycle at 25', auto horn at 10'	—
90	Loud	Lawn mower, noisy city street	Full symphony band, noisy factory
80	Moderately loud	Diesel truck at 50'	Garbage disposal, dishwasher
70	Average	—	Face-to-face conversation, vacuum cleaner, printers, and copiers
60	Moderately quiet	Air conditioning condenser at 15', auto traffic near an interstate highway	Normal conversation, general office
50	Quiet	Large transformer at 50'	—
40	Very quiet	Bird calls	Private office, soft radio music
30	Extremely quiet	Quiet residential neighborhood	Average residence
20	Nearly silent	Rustling leaves	Quiet theater, whisper
10	Just audible	—	Human breathing
0	Threshold of human hearing	—	One's own heartbeat in a silent room

transmission class (STC). An STC value of 20 to 25 indicates that even normal speech can be easily understood in an adjoining room. On the other hand, an STC value of 50 to 60 indicates that loud sounds will be heard only faintly or not at all. Acceptable STC ratings range from approximately 50 to 65.

- *Articulation class (AC)* — The rating of a ceiling's ability to achieve normal privacy in open office spaces by absorbing noise reflected at an angle off the ceiling into adjacent areas (cubicles). According to ASTM International *E1110* and *E1111* standards, the generally accepted AC ratings for normal privacy in open-plan offices is a minimum of 170, with 190 to 210 preferred.

- *Ceiling attenuation class (CAC)* — The rating of a ceiling's efficiency as a barrier to airborne sound transmission between adjacent work areas, where sound can penetrate **plenum** spaces and travel to other spaces. CAC is stated as a minimum value. Per *ASTM E1264*, CAC minimum 25 is acceptable in an open-plan office, while a rating of CAC minimum 35 to 40 is preferred for closed offices.

- *Absorption* — The energy of sound waves being taken in (entering) and absorbed by a surface of any material rather than being bounced off or reflected.

Plenum: A chamber or container for moving air under a slight pressure. In commercial construction, the area between the suspended ceiling and the floor or roof above is often used as the HVAC return air plenum.

1.1.0   Ceiling Systems

A wide variety of suspended ceiling systems is available, with each system being somewhat different from the others. All use the same basic materials, but their appearances are completely different.

The focus of this module is on the following ceiling systems:

- Exposed grid systems
- Metal-pan systems
- Direct-hung concealed grid systems
- Integrated ceiling systems
- Luminous ceiling systems
- Suspended drywall ceiling system
- Special ceiling systems

Also covered in this module is background information relevant to acoustics and acoustical ceilings, including information on the propagation of sound waves, acoustical ceiling product terminology, and ceiling-related drawings.

1.1.1 Exposed Grid Systems

An exposed grid system is a suspension system for lay-in ceiling tiles (*Figure 2*). The factory-finished supporting members are exposed to view.

Figure 2 Exposed grid system.
Source: Lukassek/Shutterstock

1.1.2 Metal-Pan Systems

The metal-pan system resembles a conventional exposed grid ceiling system except that metal panels or pans are used in place of the conventional sound-absorbing tile. In some cases, the panels or pans are snapped into place from below rather than being laid-in from above the ceiling frame.

1.1.3 Direct-Hung Systems

A direct-hung ceiling system is used if the grid system is to be concealed from view. A mechanical clip or tongue-and-groove joint is used to connect the tiles together. The tiles are then tied to the suspended grid.

1.1.4 Integrated Ceiling Systems

As suggested by its name, the integrated ceiling system incorporates the lighting and/or an air supply **diffuser** as part of the overall ceiling system, as shown in *Figure 3* and *Figure 4*.

Diffuser: An attachment for duct openings in air distribution systems that distributes the air in wide flow patterns. In lighting systems, it is an attachment used to redirect or scatter the light from a light source.

Figure 3 Integrated grid system.
Source: Art Noppawat/Shutterstock

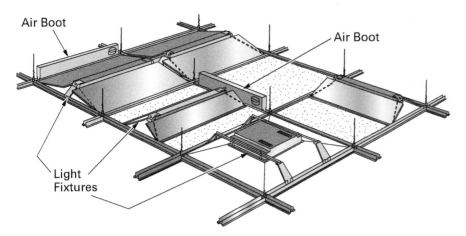

Figure 4 Integrated ceiling schematic.

Some manufacturers of ceiling materials offer specialty ceiling designs. Many integrated ceilings can be used to create distinctive architectural appearances and interior artistic designs. Complete ceiling systems offer dozens of options, both functional and visual.

The functional aspect allows for enhanced lighting and lighting effects. Mechanically, it can incorporate the air supply system through spaced air supply diffusers and return air grilles, all of which have been designed to go beyond function to enhance the artistic appearance of the ceiling.

Integrated ceiling systems are available in units called modules. The common sizes are 30" × 60" and 60" × 60". The dimensions refer to the spacing of the main runners and cross tees.

1.1.5 Luminous Ceiling Systems

Whereas light sources are incorporated into an integrated ceiling, luminous ceiling systems are the light source. Unlike other ceiling systems, in which the light sources interrupt the surface of the ceiling, the surface of luminous ceilings are a continuous light source (only interrupted by the ceiling grid if an exposed grid is used). The choice to make the entire ceiling a light source is often made to provide dispersed light—rather than direct light which tends to create a glare—to an area that needs to be well lit, such as a museum or office. Luminous ceiling systems (*Figure 5*) are available in many styles, such as exposed grid systems with drop-in plastic light diffusers or an aluminum or wood framework with translucent acrylic light diffusers.

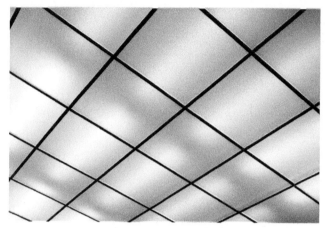

Figure 5 Luminous ceiling system.
Source: Alex Veresovich/Shutterstock

Fluorescent or LED fixtures are generally installed above the translucent diffusers. Standard modules of 2' × 2' up to sizes of 5' × 5' are available. It is also possible to purchase custom sizes for special fit conditions. There are two types of luminous ceilings: standard and nonstandard. Standard systems are, as their name indicates, those that are available in a series of standard sizes and patterns. Nonstandard systems differ in that they deviate from the normal spacing of main supports and may include unusual tile sizes, shapes, and configurations.

All surfaces in the luminous space, including pipes, ductwork, ceilings, and walls, are painted with a 75% to 90% reflective matte white finish. Any surfaces in this area that might tend to flake, such as fireproofing and insulation, should receive an approved hard surface coating prior to painting to prevent flaking onto the ceiling below.

1.1.6 Suspended Drywall Ceiling System

The suspended drywall system is used when it is desirable or specified to use a drywall finish or drywall backing for an acoustical panel ceiling.

1.1.7 Special Ceiling Systems

Ordinary acoustical ceilings with an exposed grid and white, mineral fiber panels are so familiar that they likely go unnoticed by most. On the other hand, some ceilings are indeed special.

This category of eye-catching ceiling systems includes (but is not limited to) those covered by wood or metal tiles or panels, arranged in soothing linear patterns, composed of neatly arranged baffles or blades, spotted with floating acoustical islands, reflecting back at you like a pond, and formed by mesmerizing curves (*Figure 6*).

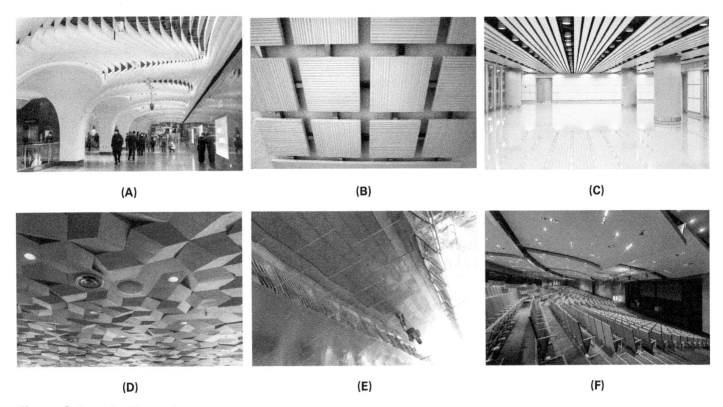

(A)   **(B)**   **(C)**

(D)   **(E)**   **(F)**

Figure 6 Special ceiling systems.
Sources: (A) Cynthia Lee/Alamy; (B) David Ausserhofer/Intro, imageBROKER.com GmbH & Co. KG/Alamy; (C) yxm2008/Shutterstock; (D) Askar Karimullin/Alamy; (E) Benjamin Marcus/Alamy; (F) bialasiewicz/123RF

Metallic Ceilings

The malleability of metal allows it to be formed into a wide variety of shapes and styles. Metallic ceilings are also desirable because they reflect rather than absorb heat, making them energy efficient. Additional benefits include ease of maintenance, durability, recyclability, and fire-resistance. Sound-absorbing backing material is installed in metallic ceiling systems to improve acoustical quality.

Pre-manufactured metallic ceiling tiles are normally produced in the standard 2' × 2' and 2' × 4' sizes and may be of the lay-in or surface mount installation type. Custom-made metallic ceiling tiles can be ordered from different manufacturers in a variety of sizes, shapes, and surface patterns.

Linear/Planar Ceilings

Linear (or planar) ceiling systems allow designers to create a sense of visual flow and focus as the lines direct the viewer's eyes. Long planks can be used to achieve a smooth, continuous design or a variety of lengths and dimensions can be used to create interesting patterns.

The uniformity and length of linear tiles speeds up the installation process. The smooth minimalist appearance of the linear tiles makes them compatible with a variety of building exteriors and, therefore, a popular choice for exterior ceilings.

Wood lends itself well as a material for use in linear ceilings because it is relatively easy to cut into uniform, linear sections. Manufactured wood linear ceiling panels are made to lay into or mount to proprietary grid systems. Wood panels can also be surface mounted and given soundproofing, in addition to the natural acoustics of the wood, through the application of insulative backing material.

Reflective Ceilings

When designers want to increase the perceived space within a structure without increasing the height of the ceiling, they might incorporate a reflective ceiling. Reflective ceilings return the existing light within a space and intensify the mood of the lighting.

Metals with high reflective properties and glass (mirrors) have traditionally been used to make reflective ceiling tiles. A glossy coating can be applied to *acoustic stretch ceiling systems* (lightweight fabric stretched between a frame with sound-absorbent membrane) to create a highly reflective result.

Baffle and Blade Ceilings

Baffles are free-hanging panels that obstruct, trap, or redirect the flow of sound. Like other acoustical ceilings, baffle systems are suspended from a support grid. The quantity, size, and perforations in the baffles provide sound absorption.

Blades, slats rotated with the narrow side facing, are often used to create open cell linear ceiling designs by leaving gaps between the blades. Blades provide sound absorption, which is dependent on the size of the blades and their spacing. The spacing of the panels also determines the degree to which the plenum or mechanicals will be visible.

Both baffles and blades are highly customizable in terms of their size, shape, material, and color.

Island/Cloud Ceilings

Similar to baffle/blade systems, acoustical islands and clouds utilize sound-absorbing panels suspended from grid systems, which allow some of the mechanical systems and structure above the panels to show. Islands and clouds are both made from sound absorbing materials that can be produced in a variety of colors and shapes, including curves and waves. The difference between the two is that islands are larger and cover more area than clouds. Islands and clouds are sometimes used in combination with baffles to increase the sound absorption of the system.

Look Up

The next time you are out, look at the ceilings in the stores, theaters, malls, and other buildings you enter. You will see that the available styles, designs, materials, and color schemes are more numerous than you ever imagined. The one thing most of them have in common is that they are some form of suspended ceiling using panels set into or attached to a framework.

Sources: (left) gerenme/Getty Images; (right) Pavel L Photo and Video/Shutterstock

1.2.0   Ceiling System Components

A variety of components are combined to create the ceiling grid. Each type of system and each manufacturer's components may be slightly different. Always refer to the manufacturer's instructions before installing a ceiling grid. The grid, composed of main runners, cross runners, and wall angles, is commonly made from light-gauge metal members.

1.2.1 Exposed Grid Components

For an exposed grid suspended ceiling, a light-gauge metal grid is hung by wires attached to the original ceiling or structural members. Tiles that usually measure 2' × 2' or 2' × 4' are then placed in the frames of the metal grid. Exposed grid systems are constructed using the components and materials described as follows and shown in *Figure 7*.

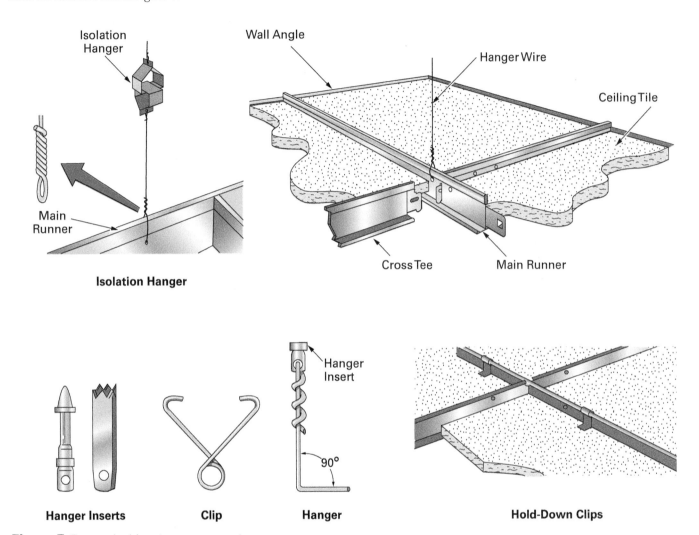

Figure 7 Exposed grid system components.

- *Main runners* — Primary support members of the grid system for all types of suspended ceiling systems. They are 12' in length and are usually constructed in the form of an inverted T. When it is necessary to lengthen the main runners, they are usually spliced together using extension inserts; however, the method of splicing may vary with the type of system being used.

- *Cross runners (cross ties or cross tees)* — Inserted into the main runners at right angles and spaced an equal distance from each other, forming a complete grid system. They are held in place by either clips or automatic locking devices. Typically, they are either 2' or 4' in length and are usually constructed in the form of an inverted T. Note that 2' cross runners are only required for use when using 2' × 2' ceiling tiles.

- *Wall angles* — Installed on the walls to support the exposed grid system at the outer edges.

- *Ceiling tiles* — Tiles that are laid in place between the main runners and cross ties to provide an acoustical treatment. The acoustical tiles used in suspended ceilings stop sound reflection and reverberation by absorbing sound waves. These tiles are typically designed with numerous tiny sound traps consisting of drilled or punched holes or fissures, or a combination of both. When sound strikes the tile, it is trapped in the holes or fissures. A wide variety of ceiling tile designs, patterns, colors, facings, and sizes is available, allowing most environmental and appearance demands to be met. Tiles are typically made of glass or mineral fiber. Generally, glass-fiber tiles have a higher sound absorbency than mineral-fiber tiles. Tile facings are typically embossed vinyl in a choice of patterns such as **fissured**, pebbled, or **striated**. The specific ceiling tiles used must be compatible with the ceiling suspension system due to variations in manufacturers' standards.

- *Hanger inserts and clips* — Many types of fastening devices are used to attach the grid-system hangers or wires to the ceiling or structural members located above the suspended ceiling. Screw eyes and star anchors are commonly used and may require a hammer drill for installation. Powder-actuated fasteners are commonly used when fastening to reinforced concrete. Clips are used where beams are available and are typically installed over the beam flanges. Then the hanger wires are inserted through the loops in the clips and secured.

- *Hangers* — The devices attached to the hanger inserts and used to support the main runners. The hangers can be made of No. 12 wire or heavier rod stock. Ceiling isolation hangers are also available to isolate the ceiling from noise traveling through the building structure.

- *Hold-down clips* — Used in some systems to hold the ceiling tiles in place.

- *Nails, screws, expansion anchors, and molly bolts* — Used to secure the wall angle to the wall. The specific fastener used depends on the wall construction and material.

Fissured: A ceiling-panel or ceiling-tile surface design that has the appearance of splits or cracks.

Striated: A ceiling-panel or ceiling-tile surface design that has the appearance of fine parallel grooves.

USDA-Compliant Tiles

The United States Department of Agriculture (USDA) is responsible for ensuring that the nation's commercial supply of meat, poultry, and egg products is safe through safety guidelines and inspections. USDA-compliant ceiling tiles are designed for use in kitchens and central food-preparation areas. These tiles are washable, waterproof, and bacteria resistant.

What Is in a Name?

The terms *ceiling panel* and *ceiling tile* have specific meanings in the trade. Ceiling tiles are typically any lay-in acoustical board that is designed for use with an exposed grid system. They do not have finished edges or precise dimensional tolerances because the grid system provides the trim-out. Ceiling panels are acoustical ceiling boards, usually 12" × 12" or 12" × 24", which are nailed, cemented, or suspended by a concealed grid system. The edges are often kerfed and cut back.

1.2.2 Ceiling Panels and Tiles

Ceiling panels and tiles range in size from 12" × 12" up to 60" × 60". Various colors and designs are available. Most are fabricated from mineral fiber and glass fiber. Depending on their design and purpose, mineral-fiber panels and tiles are made with painted, plastic, aluminum, ceramic, or mineral faces. Glass-fiber panels and tiles are made with painted, film, glass cloth, and molded faces.

The three general types of tile/ceiling grid interfaces are lay-in, concealed tee, and profiled edge. Lay-in tiles are widely used and are generally the most cost-effective style. Concealed tee tiles are butt-jointed tiles that provide a monolithic ceiling design with no visible support system. Profiled edge tiles feature a wide selection of edge designs from soft-edged chamfered or curved tiles to highly articulated edges. Beveled and angular reveal-edged tiles provide a three-dimensional look in a suspended ceiling.

Many interior spaces have specific requirements for ceiling tile materials and characteristics. These can include sound control, fire resistance, thermal insulation, light reflectance, and moisture resistance (*Table 2*). Other considerations include maintenance, appearance, and cost considerations.

High-performance acoustical tiles are used to help prevent noise in open-plan and closed types of offices. The following three factors contribute to noise distractions in a workplace:

- *General office noise* — The ability of a ceiling material to absorb general office noise is measured using a value known as the noise reduction coefficient (NRC),

TABLE 2 Ceiling Panel Ratings

Rating	Meaning
NRC (Noise Reduction Coefficient)	Ability of a material to absorb general indoor noise, measured in average percentage of noise absorbed.
AC (Articulation Class)	Ability of material to absorb reflected conversational noise. A rating of 170 is generally the minimum for open plan offices.
CAC (Ceiling Attenuation Class)	Ability of material to absorb sound transmission between adjacent indoor spaces. A minimum of 25 is considered acceptable for open spaces and 35–40 is preferred for closed spaces.
Flame Spread Rating	A measure of the flammability of a material, with an index of 0–200 grouped into classes. Lower is better; a class A or class 1 rating indicates that the material is nearly incombustible (0–25 index).
Fire Resistance Assembly Rating	Degree to which an entire assembly prevents the spread of fire without losing its structural function, measured in time. The rating applies to the entire floor-ceiling or roof-ceiling assembly, not to the ceiling alone.
Thermal Insulation	When required by codes, must be lightweight. Fiberglass is preferred; if insulation is installed on non-fiberglass panels it must be R-19 with vapor barrier facing down and perpendicular to cross tees. Insulation should not be installed above fire-resistant ceiling panels.
High Humidity Resistance	Resists sagging caused by moisture and prevents growth of mold or mildew. Used in humid environments (indoor pools, bathrooms, etc.) and to control moisture from cycling HVAC systems.
High Light Reflectance (LR)	Percentage of visible light reflected. Used to maximize natural lighting in an area and reduce artificial lighting costs and environmental impacts. A rating of .83 (83% of visible light reflected) is desirable.

which is used by tile manufacturers to compare the noise absorbency of their ceiling tile products. The higher the number, the better the absorbency. The NRC measures the average percentage of noise a tile absorbs at various frequencies.

- *Reflected conversational noise that angles off ceilings into adjacent cubicles* — The ability of a ceiling material to absorb reflected conversational noise is measured using a value known as the articulation class (AC), which rates a ceiling's ability to achieve normal privacy in open office spaces by absorbing noise reflected at an angle off the ceiling into adjacent areas (cubicles). The *ASTM E1110 and E1111* standards recommend a ceiling with an AC rating above 170 for adequate privacy in an open office.

- *Sound transmission through cubicles, partitions, walls, and ceilings* — The ability of a ceiling material to absorb sound transmission is measured using a value known as the ceiling attenuation class (CAC), which rates a ceiling tile's efficiency as a barrier to airborne sound transmission between adjacent work areas, where sound can penetrate plenum spaces and carry to other spaces. The CAC is stated as a minimum value. Per *ASTM E1264*, CAC minimum 25 is acceptable in an open-plan office, while a rating of CAC minimum 35 to 40 is preferred for closed offices.

Fire-resistant ceiling tiles and support systems are specially made of materials that provide increased resistance against flame spread, smoke generation, and/or structural failure in the event of a fire. Two ratings based on ASTM International, ANSI (American National Standards Institute), and NFPA (National Fire Protection Association) standards are used to evaluate fire-resistant tiles: the flame-spread rating of the material *(ASTM E84)* and the fire-resistance rating of a ceiling assembly *(ANSI/UL 263, ASTM E119, and NFPA 251)*.

Basically, the flame-spread rating of a ceiling material is the relative rate at which a flame will spread over the surface of the material. This rate is compared against a rating of 0 (highest rating) for fiber-cement board and a rating of 100 for red oak. Class A ceilings have flame-spread ratings of 25 or less, the required standard for most commercial applications. The fire-resistance rating of a ceiling assembly represents the degree to which the entire assembly, not the individual components, withstands fire and high temperatures (measured in hours). Specifically, it is an assembly's ability to prevent the spread of fire between spaces while retaining structural integrity.

Fire-Resistance-Rated Applications

Suspended ceiling systems can be used in fire-resistance-rated applications. Ratings consider factors such as resistance to fire and flame spread for both the tiles and the suspension system. However, the ceiling materials alone do not determine the fire-resistance rating. Rather, the materials and construction methods used in the entire system, including the floor/ceiling or ceiling/roof assembly, all factor into the rating determination. If a suspended ceiling is to be part of a fire-resistance-rated system, you must consult the manufacturer's product literature to determine the ceiling tile and grid that must be used to meet the rating.

Most ceiling tiles provide little thermal insulation between the space above the suspended ceiling and the work area below. When the building design or local codes require insulation above ceiling panels, special care must be taken to avoid placing too much weight on the ceiling system. When insulation is required, fiberglass ceiling panels are recommended. Most manufacturers discourage the use of insulation on mineral ceiling panels because the panels may sag as the humidity increases.

High-humidity-resistant tiles are tiles that have superior resistance to sagging caused by highly humid conditions. Many also inhibit the growth of mold or mildew that may appear in highly humid conditions. Sagging not only diminishes the attractiveness of a ceiling, but also causes ceilings to chip and soil more easily, reducing the light reflectivity of the tiles. High-humidity-resistant tiles are typically installed in humid climates or areas such as kitchens, locker rooms, shower areas, and indoor pools; buildings where the HVAC systems may be shut down for extended periods; or where the ceiling might be installed early in the construction process before the building is fully enclosed.

Ceiling tiles with a high light reflectance (LR) value of 0.83 or greater per *ASTM E1477* help increase effective lighting levels and reduce light fixture costs and energy consumption, especially with indirect lighting systems. Their use also helps to reduce eyestrain. These tiles typically have soil-resistant surfaces that stay cleaner longer than standard ceilings, resulting in a much lower loss in light reflectance over time.

Detailed information regarding the various tile characteristics described in this section can normally be found in ceiling manufacturers' product catalogs and literature. When selecting tiles for a particular application, it is best to follow the manufacturer's recommendations.

1.2.3 Metal-Pan System Components

The metal-pan system is similar to the conventional suspended acoustical ceiling system except that metal panels, or pans, are used in place of the conventional sound-absorbing tile (*Figure 8*). In some cases, the metal pans are snapped in from the bottom of the grid.

The pans are made of steel or aluminum and are generally painted white; however, other colors are available by special order. Pans are also available in a variety of surface patterns. Metal-pan ceiling systems are effective for sound absorption. They are durable and easily cleaned and disinfected. In addition, the finished ceiling has little or no tendency to have sagging joint lines or drooping corners. The metal pans are die-stamped and have crimped edges, which snap into the spring-locking main runner and provide a flush ceiling.

The tools, room layout, and installation of hanger inserts, hangers, and wall angle for the metal-pan system are basically the same as for the conventional exposed grid suspended ceiling.

1.2.4 Direct-Hung System Components

A concealed grid system is advantageous if the support runners need to be hidden from view, resulting in a ceiling that is not broken by the pattern of the runners (*Figure 9*).

The panels used for this system are similar in composition to conventional panels but are manufactured with a kerf on all four edges. Kerfed and rabbeted

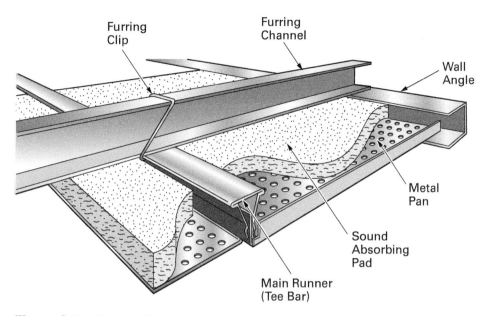

Figure 8 Metal-pan ceiling components.

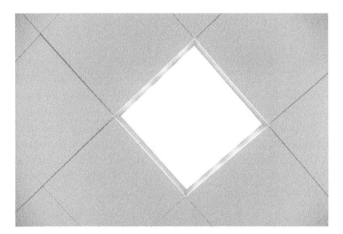

Figure 9 Concealed grid system.
Source: Ansario/Shutterstock

12" × 12" and 12" × 24" panels are used with this system. Splines are inserted in the kerfs to tie the panels together. Panels of various colors and finishes are available. Refer to *Figure 10* for a diagram of the components in a typical concealed grid system.

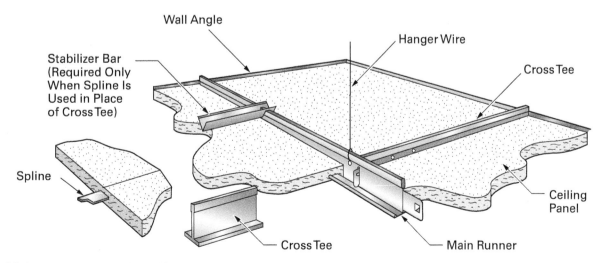

Figure 10 Direct-hung concealed grid system components.

1.3.0   Material Handling and Storage

Ceiling panels or tiles, and in some cases the tracks and runners, are exposed to view. To prevent damage, they must be handled and stored carefully. In addition, working with ceiling materials often presents safety hazards.

1.3.1 Handling and Storing Ceiling Materials

Finish ceiling materials should be stored in their original unopened packages and be protected from damage and exposure to the elements. Materials in their unopened packages can be easily moved to their installation location. The conditions where the materials are stored should be as close as possible to the place where they will be installed. Ensure there is proper support for the ceiling materials being placed. Materials should be stored at the jobsite for a minimal amount of time before being installed. Long-term storage should be avoided. Other considerations for proper handling and storage are as follows:

- Excess humidity during storage can cause expansion of material and possible warp, sag, or poor fit after installation.
- Chemical changes in the mat and/or coatings can be aggravated by excess humidity and cause discoloration during storage, even in unopened cartons.
- Cartons should be removed from pallets and stringers to prevent distortion of material.

1.3.2 Safely Working with Ceiling Materials

When installing grid support members or the panels or tiles themselves, carpenters typically work from ladders, movable scaffold, or lifts, depending on the height of the ceiling. Refer to the NCCER Module 00101, *Basic Safety (Construction Site Safety Orientation)* for specific safety information relating to ladders and lifts. A common type of movable scaffold used for ceiling projects is called a baker's scaffold. Baker's scaffolds are short and lightweight scaffolds with wheels that allow scaffolds to roll easily. Safety guidelines for using a baker's scaffold include the following:

- Inspect the scaffold before each use for defects or damage.
- Do not stand on or attach any equipment to cross braces or diagonal braces.
- Do not place boxes or ladders on a scaffold to increase your reach or height.
- Do not sit or stand on guardrails. Ensure that all guardrails are secured in place on all four sides.
- Never ride on a moving scaffold.
- Do not attempt to move a scaffold by applying a pushing or pulling force at or near the top of the scaffold.
- When hoisting material up to a scaffold platform, ensure the scaffold is attached to a permanent structure to keep the scaffold from tipping.
- Workloads on the scaffold must not exceed the capacity of the lowest-rated scaffold component.
- When working around a scaffold, always wear a hard hat, safety glasses, work gloves, and steel-toe boots.

In addition to standard hand-tool and power-tool safety guidelines that should be followed, eye protection is especially important when installing hanger wire. The loose ends of hanger wire are commonly at eye level when working from a baker's scaffold. Carefully handle the cut ends of wall angle and other support members. The cut ends of these items are very sharp and can cause serious injury.

Using Stilts to Install Ceilings

While stilts are permitted in certain areas, they are not permitted in all jurisdictions. Always refer to the specific safety regulations in effect in the area in which you are working before using stilts.

1.0.0 Section Review

1. Integrated ceiling systems incorporate _____.
   a. both ceiling panels and ceiling tiles
   b. fire-rated panels and sprinkler systems
   c. lighting and/or air supply diffusers
   d. both exposed and concealed grid systems

2. True or False: A typical ceiling grid consists of main runners, cross runners, and wall angles.
   a. True
   b. False

3. Finished ceiling materials should be stored _____.
   a. on pallets
   b. in their original unopened packages
   c. at a constant 60°F
   d. on edge to prevent warping

2.0.0 Reflected Ceiling Plan

Objective

Interpret a reflected ceiling plan.

a. Interpret the layout information.

b. Interpret the mechanical, electrical, and plumbing (MEP) locations.

Performance Task

1. Estimate the quantities of materials needed to install a lay-in suspended ceiling system in a typical room from an instructor-supplied drawing.

Some large construction jobs may have a set of reflected ceiling plans. These plans show the details of the ceiling as though it were reflected onto the floor (*Figure 11*). This view shows features of the ceiling while keeping those features in proper relation to the floor plan. For example, if a pipe runs from floor to ceiling in a room and is drawn in the upper left corner of the floor plan, it is also shown on the upper left corner of the reflected ceiling plan of that same room. Reflected ceiling plans show in detail how the ceiling will be constructed. The plans will indicate the following:

- Layout (direction) of the ceiling panels or tiles
- Location of the center (starting) line
- Size of the borders
- Position of the light fixtures
- Location of the air diffusers
- Location of life safety devices
- Position of strobes
- Location of exit devices
- Location of fire calls

As a rule, reflected ceiling plans also show all items that penetrate the ceiling, including the following:

- Return grilles
- Diffusers
- Sprinkler heads
- Light fixtures
- Recessed speakers for sound systems

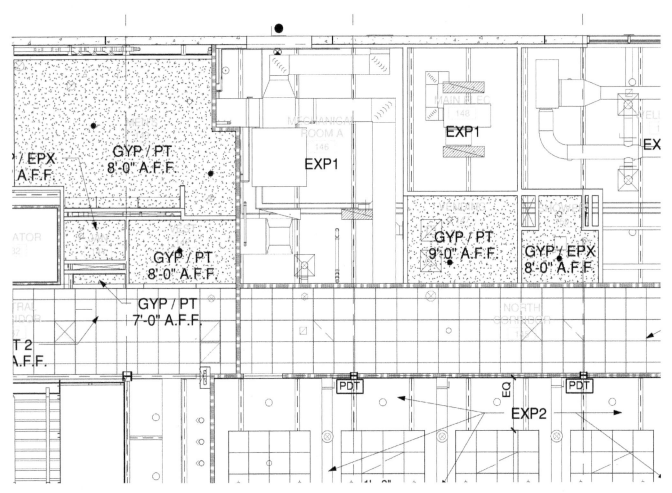

Figure 11 Example of a reflected ceiling plan.

Other information is also used when constructing a ceiling. Your employer may prepare shop drawings that show in detail just how the ceiling should be installed and also indicate the finished appearance. These drawings provide insurance against errors in the details of installation.

2.1.0 Interpreting Ceiling Plans

Pinpointing the locations of lighting fixtures, sprinkler systems, air diffusers, and other features that protrude through ceilings is essential when installing a ceiling. Other professionals working behind you depend on these items to be located accurately. In addition to the location of ceiling components found in the plans, specifications contain pertinent information including construction techniques, specifics about materials, and dimensions.

To avoid mistakes and/or omissions when installing ceilings, an organized and systematic approach should be used for reading the related construction drawings. The following is a general procedure for reading construction drawings:

Step 1   Check the room schedule on the construction drawings.
- Identify the type of material to be used.
- Locate the protrusions into the ceiling.

Step 2   Locate the room on the floor plan.
- Find the room dimensions. If none are found, locate the drawing scale.
- Using the given scale, determine the dimensions of the room.

Step 3   Check to see if there is a reflected ceiling plan. If no reflected ceiling plan is found, check to see if a shop drawing is included.

Step 4   Be sure the construction drawings are the final revised set.

- Construction drawings are often revised several times before the ceiling is ready to be installed. To be sure the construction drawings are the final revised set, check the revision block date against the work order to see if they are the same.
- If work that has already been done is not reflected on the construction drawings, chances are the construction drawings are not the final revised set. This could have a significant impact on the ceiling installation.

Step 5   Read the specifications and general notes.

- Be sure the ceiling to be installed is the same as that listed in the specifications.
- If the job conditions do not agree with what is shown in the specifications and/ or plans, call your supervisor and ask for instructions on how to proceed.

Step 6   Check the mechanical and electrical plans prior to the layout of the ceiling.

- On the mechanical plans, locate the air diffusers (HVAC air supply outlets), return grilles, ducts, and sprinkler heads.
- On the electrical plans, locate the light fixtures, fans, and other ceiling protrusions.
- Make sure to cross-reference both plans for items that may be in the same location.

2.2.0   Interpreting the Mechanical, Electrical, and Plumbing (MEP) Drawings

Along with reading and interpreting the reflected ceiling plan, reference the mechanical, electrical, and plumbing (MEP) drawings to ensure that all ceiling penetrations are accounted for. The MEP drawings will indicate where plumbing and electrical risers penetrate the ceiling, and will show the routing of mechanical equipment. *Figure 12A* and *Figure 12B* show the mechanical, electrical, and plumbing drawings for similar areas of the same set of construction drawings.

Plenum Ceilings

The systems that provide heating and cooling for most commercial buildings are forced-air systems. Blower fans are used to circulate the air. The blower draws air from the space to be conditioned and then forces the air over a heat exchanger, which cools or heats the air. In a cooling system, for example, the air is forced over an evaporator coil that has very cold refrigerant flowing through it. The heat in the air is transferred to the refrigerant, so the air that comes out the other side of the evaporator coil is cold. In homes, the air is delivered to the conditioned space and returned to the air conditioning/heating system through ductwork that is usually made of sheet metal. In commercial buildings with suspended ceilings, the space between the ceiling and the overhead decking is often used as the return air plenum. It is often called an open plenum. (A plenum is a sealed chamber at the inlet or outlet of an air handler.) This approach saves money by eliminating about half the cost of materials and labor associated with ductwork.

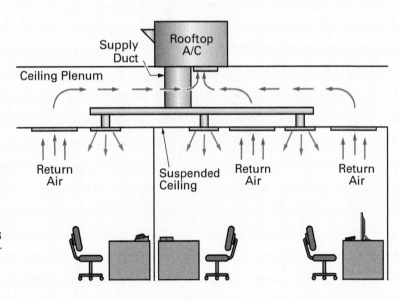

One thing to keep in mind is that anything in the plenum space (electrical or telecommunications cable, for example) must be specifically rated for plenum use in order to meet fire-resistance ratings. Plastic sheathing used on standard cables gives off toxic fumes when burned. Plenum-rated cable uses nontoxic sheathing.

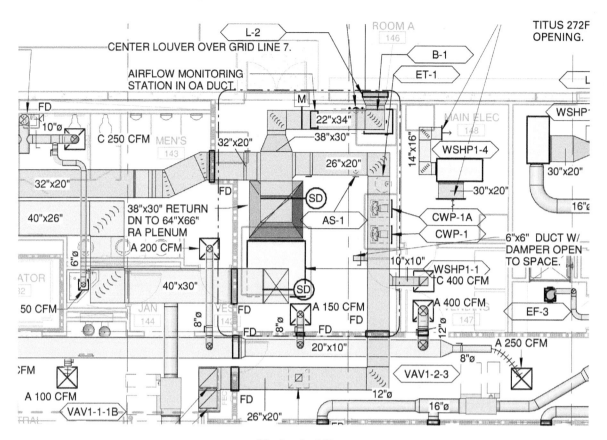

Mechanical Plan

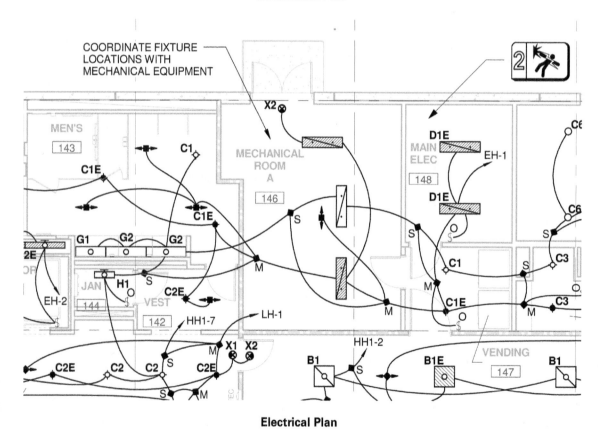

Electrical Plan

Figure 12A Mechanical, electrical, and plumbing drawings. (1 of 2)

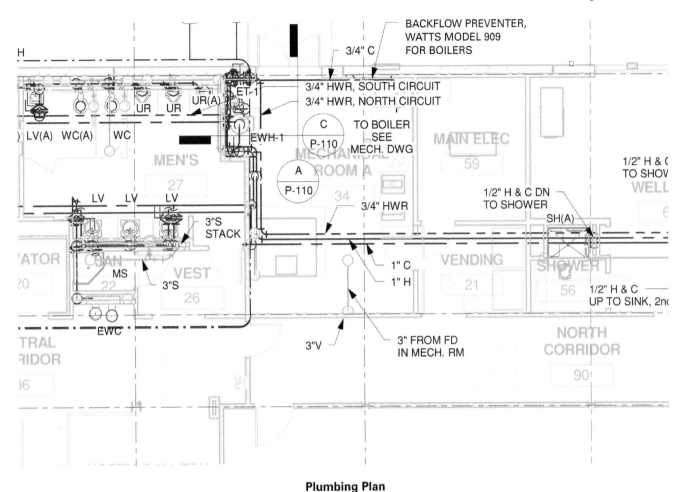

Plumbing Plan

Figure 12B Mechanical, electrical, and plumbing drawings. (2 of 2)

2.0.0 Section Review

1. To be sure that the construction drawings you are using are the final revised set, you should _____.
    a. verify the revision block date
    b. ask your supervisor
    c. call the architect's office
    d. compare them to the specifications

2. MEP drawings indicate _____.
    a. the direction and location of main runners
    b. where open plenums should be installed
    c. datum lines for ceiling elevations
    d. where plumbing and electrical risers penetrate the ceiling

3.0.0 Laying Out and Installing Suspended Ceiling Systems

Performance Tasks

2. Establish a level line at ceiling level, such as is required when installing the wall angle for a suspended ceiling.
3. Lay out and install a lay-in suspended ceiling system according to an instructor-supplied drawing.

Objective

Describe the key considerations, methods, and best practices relating to ceiling installation.

a. Explain seismic considerations for ceilings.
b. Identify the layout and takeoff procedures to install a suspended ceiling system.
c. Identify the tools and equipment to lay out and install a suspended ceiling system.
d. Identify the installation methods and procedures for a suspended ceiling system.

A professional installer must understand a variety of ceiling systems, including their components, acoustical properties, and installation techniques. In addition, the installer should be familiar with the following aspects of suspended ceilings:

- The risk of structural failure due to seismic activity is a concern, especially in earthquake-prone areas. Special building codes apply to the materials and design of ceiling systems depending on the seismic risk category.

- Before a ceiling installation job begins, the cost of materials must be accurately estimated for bidding and/or budgeting purposes.

- To maintain an attractive appearance and retain acoustical properties, ceilings need to be cleaned periodically. Tiles and panels may have specific cleaning instructions.

- Installers should follow plan and manufacturer's guidelines to ensure a superior installation.

3.1.0   Seismic Considerations for Ceilings

Although important in their function, modern ceilings are generally not part of a building's structural system. Ceilings today are used to make rooms and buildings more livable. The most desirable features of a ceiling are those that improve the acoustic properties of an area and those that make it easier to heat, cool, and light. Although ceilings are not structurally significant, the *International Building Code®* (*IBC®*) recognizes that ceiling failures due to seismic activity can make an area unusable.

Although seismic activity is of special importance in earthquake-prone areas, such as the Pacific Northwest region of the United States, it is also an issue in other areas. The *IBC®* uses three factors that place more than half of the US in areas at risk for seismic activity. The factors the *IBC®* uses for determining a building's seismic risk are as follows:

- *Ground movement* — All geographic areas have the potential for ground movement, even if the movement is unlikely to be felt. The primary factor is the amount of potential ground movement. In the US, the state with the potential for the most ground movement is California, while the ones with the least potential are Florida and Texas.

- *Soil classification* — Soil classifications are based on the soil type at the building site to a point 100" below the surface. For seismic purposes, there are six soil classifications, labeled A through F. Class A is solid rock, which is considered the most stable. Class F soil has no positive construction traits and is unusable for building. It is unlikely that a site would be classified at the extremes of A or F. Most sites will fall somewhere in between.

- *Building use/occupancy category* — Risk categories based on occupancy of a structure are divided into four groups. Category I buildings have low occupancy, such as small agricultural buildings, and they represent a low risk of loss of human life in the event of a failure. Category II consists of all buildings and other structures not listed in the other three categories. Structures that represent substantial hazard to human life in the event of structural failure fall into Category III. Category IV structures are designated as essential facilities, meaning they are required to maintain the functionality of structures in the other three categories.

3.1.1 Seismic Design Categories

Based on factors of potential ground movement, soil type, and building use (risk categories), a seismic design category (SDC) has been established. The categories range from A (lowest seismic risk) to F (highest risk). Based on the SDC assigned to a geographical region in which a structure is located, certain building codes apply to the design and installation of ceiling systems. Buildings in categories A and B use the standard suspended ceiling requirements described in this module. Their ceilings are anchored to walls and suspended from the upper decks using standard hardware. Those buildings in category C should have unrestrained ceilings that are free-floating to allow for movement of up to 12" and 45° without damage. Buildings in categories D, E, and F should have restrained ceilings that are reinforced with stabilizing bars, heavier gauge wire, and multiple strands of wire. *Figure 13* is an earthquake hazard map showing seismic risk across the United States.

The ceilings described in categories C through F are more expensive than those in categories A and B. Unrestrained ceilings (used in category C buildings) are not anchored to walls, so they are extremely difficult to install without

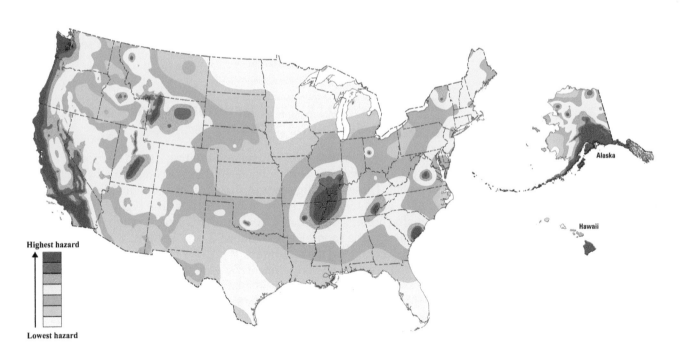

Figure 13 Map of seismic hazard in the US.
Source: U.S. Geological Survey

a fixed reference. Unrestrained ceilings cost about 50% more than standard suspended ceilings—mostly in installation costs. Unrestrained ceilings have the following features:

- The ceiling system grid is not attached to the wall in any way.
- There are no perimeter wires.
- The wall molding must be at least $\frac{7}{8}$" in width.
- The clearance between the ceiling and walls is at least $\frac{3}{8}$" at all points.
- Grids joints must overlap at least $\frac{3}{8}$".
- The ends of the main and cross tees, which are free-floating, must be tied together to prevent spreading.

Restrained ceilings (used in categories D, E, and F) are anchored to at least two adjacent walls and require the use of heavy gauge hardware, as well as a grid system. These ceilings usually cost twice as much as a standard suspension ceiling, with materials accounting for most of the increased cost. Restrained ceilings have the following characteristics:

- A heavy-duty ceiling grid system is required.
- The grid must be attached to two adjacent walls, and the clearance between the ceiling and the opposite wall must be $\frac{3}{4}$".
- The wall molding must be at least 2" wide.
- The ceiling perimeter must have support wires that are spaced not more than 8' from the wall.
- Ceilings with an area of less than 1,000 ft^2 must use heavy-duty hardware for equipment mounted through the ceiling, such as light fixtures and sprinklers.
- Ceilings with an area of more than 1,000 ft^2 must have additional bracing.
- Ceilings with an area of more than 2,500 ft^2 must have seismic separation joints.
- Cable trays and electrical conduits must have their own supports and braces.

Requirements for acoustical tile and lay-in panel ceilings based on seismic design categories are more complex than the overview provided here. For detailed codes and requirements, see the *Additional Resources* section at the end of the book.

3.2.0 Laying Out and Estimating Materials for a Suspended Ceiling

The estimate of materials for a suspended ceiling should be based on the ceiling plan provided with the construction drawings or on a scaled sketch of the ceiling layout. These drawings should show the direction and location of the main runners, cross tees, light tiles, and border tiles. In a typical suspended ceiling, the main runners are spaced 4' apart and are usually run parallel with the long dimension of the room. For a standard 2' × 4' pattern, 4' cross tees are spaced 2' apart between the main runners. If a 2' × 2' pattern is used, 2' cross tees are installed between the midpoints of the 4' cross tees.

If no ceiling plan or sketch is available, a craft professional may need to make one to determine the required quantity of materials. A sketch can be made following the steps shown in Section 3.4.1, *Exposed Grid Systems. Figure 14* shows an example of a sketched ceiling plan.

From the ceiling plan or sketch, determine the number of pieces required for the wall angle, main runners, cross tees, and ceiling tiles. Normally, main runners come in 12' lengths, cross tees in 4' lengths, and wall angle in 10' lengths. The following calculations are based on the example room shown in *Figure 14*:

- Using the ceiling plan or sketch, find the number of main-runner sections needed. For our example room, six main-runner sections are required. This

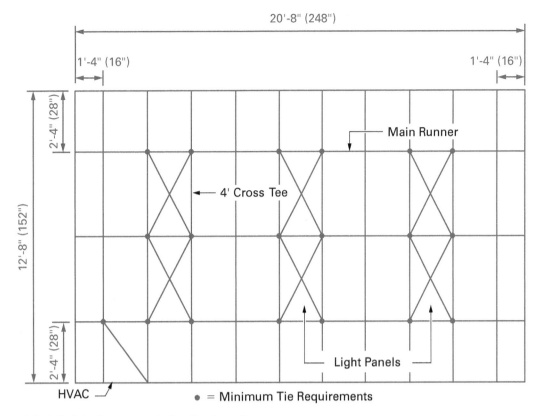

Figure 14 Completed sketch of a suspended ceiling layout.

is because main runners are made 12' in length and no more than two pieces can be cut from any one 12' runner.

$$3 \text{ main runners} \times 20'8" = 62.001'$$
$$62.001' \div 12' = 5.167 \text{ rounded to 6 lengths}$$

- Find the number of 4' cross tees. For our example room, 40 cross tees are required. Note that the border cross tees must be cut from full-length cross tees.

$$10 \text{ cross tees per row} \times 4 \text{ rows} = 40 \text{ cross tees}$$

  ○ If using 2' × 2' panels or tiles, find the number of 2' cross tees. A 2' × 2' grid is made by installing 2' cross tees between the midpoints of the 4' cross tees. The number of 2' cross tees required for our example room is 44 (11 per row × 4 rows).

- Find the number of sections of wall angle needed. Divide the perimeter of the room by 10' [perimeter = (2 × length) + (2 × width)]. For our example room, the perimeter is 66'-8" (66.667'). Therefore, seven sections are needed (66.667' ÷ 10' = 6.667, or 7 when rounded off).

$$\text{Perimeter} = (2 \times 152") + (2 \times 248")$$
$$\text{Perimeter} = 66'8" = 66.667'$$
$$66.667' \div 10' = 6.667 = 7 \text{ sections}$$

- Find the number of ceiling panels or tiles. One method is to count the total number of ceiling tiles shown on the ceiling plan or sketch. Note that each border tile requires a full-size ceiling tile. Also, subtract one tile for each lighting fixture installed in the ceiling. Assuming the use of 2' × 4' ceiling tiles and six light fixtures, our example ceiling would require 38 tiles.
- Find the approximate number of hanger wires and hangers needed. Assume one hanger device and wire for about every 4' of main runner. For our example ceiling, approximately 16 hangers are required (62.001' ÷ 4' = 15.5, or 16 when rounded off). Multiply the number of hangers needed by the required length of each hanger wire to find the total linear feet of hanger wire needed.

NOTE

Another method for estimating the number of ceiling tiles is to determine the total area (in square feet) of the ceiling by multiplying the length by the width (area = length × width). Then, divide the total ceiling area by the coverage (in square feet) printed on the carton of the tiles intended for use. If using tiles that cover 64 ft^2 per carton, our example room ceiling would require 4.09 cartons of ceiling tiles (12.667' × 20.667' = 261.789' ÷ 64 = 4.09, or four cartons plus one extra tile).

3.2.1 Alternate Method for Laying Out a Suspended Ceiling Grid System

Another common method for laying out the grid system for a suspended ceiling is given here:

Step 1 Locate the room centerline parallel to the long dimension of the room and draw it on the ceiling sketch.

Step 2 Beginning at the centerline and going toward each side wall, mark off 4' intervals on the sketch. If more than a 2' space remains between the last mark and the side wall, locate the main runners at these marks. If less than a 2' space remains between the last mark and the side wall, locate the main runners at 4' intervals beginning 2' on either side of the centerline. This procedure provides for symmetrical border tiles of the largest possible size. Remember to consider the locations of light fixtures and air diffusers in the room.

Step 3 Locate the 4' cross tees by drawing lines 2' on center at right angles to the main runners. To obtain border tiles of equal size, begin at the center of the room using the same procedure as in Step 2.

Step 4 If using a 2' × 2' grid pattern, locate the 2' cross tees by bisecting each 2' × 4' module.

Step 5 Estimate the materials for the grid system, using the information shown on the ceiling sketch in the same manner as described previously.

3.2.2 Establishing Room Centerlines

For a grid system to be installed square within a room, it is necessary to lay out two centerlines (north-south and east-west) for the room. When correctly laid out, these centerlines will intersect at right angles (90 degrees) at the exact center of the room. If the ceiling of the room located above the proposed suspended ceiling is solid and flat, such as with a plaster or drywall ceiling, then the centerlines can be laid out on the ceiling. If the ceiling is not flat, such as an open ceiling with I-beams or joists, the centerlines can be laid out on the floor. In either case, the centerlines that are laid out on the ceiling or floor can be transferred down from the ceiling, or up from the floor, to the level of the suspended ceiling by use of a plumb bob and **dry lines**. The following procedure describes one common method for establishing the centerlines in a rectangular room:

Dry lines: A string line suspended from two points and used as a guideline when installing a suspended ceiling.

Step 1 Measure and mark the exact center of one of the short walls in the room. Repeat the procedure at the other short wall.

Step 2 Snap a chalk line on the ceiling (or floor) between these two marks. (See *Figure 15*, chalk line A-B). This is the first centerline.

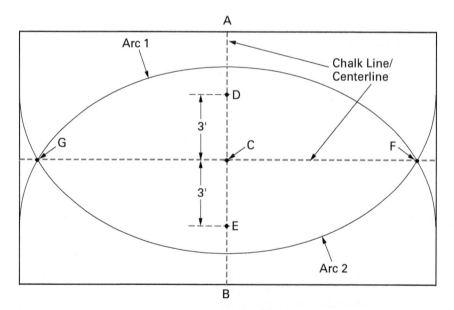

Figure 15 Method for laying out the centerlines of a room.

Step 3   Measure the length of the room along the chalk line. Find its center, and then place a mark on the chalk line at this point (point C).

Step 4   From point C, measure a minimum of 3' in both directions along the chalk line, then place a mark on the chalk line at these points (points D and E).

Step 5   Drive a nail at point D and attach a string to the nail. Extend the string to the side wall so that it is perpendicular to the chalk line, then attach a pencil to the line.

Step 6   Making sure to keep the string taut, draw an arc on the ceiling (floor) from the wall, across the chalk line, to the opposite wall (arc 1).

Step 7   Repeat Steps 5 and 6, starting at point E on the chalk line and draw another arc (arc 2).

Step 8   Mark the intersecting points of arcs 1 and 2 on both sides of the chalk line (points F and G).

Step 9   Snap a chalk line from wall to wall on the ceiling (or floor) that passes through points F and G. If done correctly, you should now have two centerlines perpendicular to each other that cross in the exact center of the ceiling (floor).

3.3.0   Suspended Ceiling Tools and Equipment

Depending on the type of suspended ceiling being installed, a variety of tools and leveling equipment is needed. Reviewing the NCCER Module 00101, *Basic Safety (Construction Site Safety Orientation)* will refresh your knowledge of safety guidelines and considerations for working on a construction project.

3.3.1 General Tools

Preparation for installing a suspended ceiling includes gathering the necessary tools and equipment. Having the right equipment ready before installation begins will set up the rest of the process for success. The following tools are required when installing an exposed grid ceiling system:

- Aviation snips (tin snips)
- Clamping pliers or vise grips with plastic or rubber corners
- Clamps and brackets
- Chalk line
- Dry line
- 50' or 100' tape measure
- Hammer
- Awl
- Keyhole saw
- Lath nippers
- Magnetic punch
- Scribe or compass
- Plumb bob
- 9" lineman's pliers
- Straightedge for cutting
- Board
- Tile knife
- Ladders
- Laser level
- Bubble level
- Pop-rivet gun
- Powder-actuated tool
- Scaffold
- Special dies for cutting suspension members
- Whitney punch

Layout Equipment

Most of the work of installing a ceiling will be done at ceiling level. Ladders, portable scaffolding, scissor lifts, or drywall stilts will all get the job done. Several factors such as the height of the ceiling, the complexity of the job, the projected timeline for completion, and the budget will impact your choice of access equipment.

Proper installation of suspended ceilings begins with taking accurate measurements. You will need a tape measure to find the center of the walls in the room, and to measure spacing of the main runners and cross runners.

Cutting and Fastening Equipment

Lineman's (electrician's) pliers (*Figure 16*) are designed to cut, bend, straighten, strip, twist, grip, and otherwise manipulate electrical wire. They also happen to be a handy tool for cutting and twisting hanger wire when installing ceiling grid systems. When cross runners, also known as cross ties or cross tees, need to be cut to length to fit the main runners, either tin snips (*Figure 17*) or aviation snips can be used.

> **NOTE**
>
> Powder-actuated fasteners are not permitted to be loaded in tension (such as supporting a ceiling) in high seismic risk areas.

Figure 16 Lineman's (electrician's) pliers.
Source: Paul Nichol/Alamy

Figure 17 Tin snips.
Source: v_zaitsev/Getty Images

Rivets are fastened where the wall angle (frame that attaches all the way around the room and supports the edge of the grid) and cross runners meet as well as at the intersection of every other cross runner and main runner to stabilize the grid. A rivet tool (*Figure 18*), also referred to as a rivet gun, is used to pop the rivets through the members being attached.

Figure 18 Rivet tool.
Source: ohotnik/123RF

Figure 19 Spring clamp.
Source: Tpopova/Getty Images

Spring clamps (*Figure 19*) or ceiling grid clamps are used at the intersections of the main runners and cross runners to temporarily hold them in place as adjustments are being made to the grid. Once the grid is completely assembled and squareness has been checked, the clamps can be replaced with rivets.

A different set of fastening tools is needed when the framework is attached directly to a ceiling deck instead of being suspended. To fasten a ceiling grid to a concrete surface, either a powder-actuated fastener or hammer drill can be used. When a furring system is being installed to support drywall on a ceiling, the $\frac{1}{2}$" cold-rolled channel is attached to the ceiling deck with an impact driver.

3.3.2 Ceiling Leveling Equipment

To install suspended ceilings, carpenters use various types of leveling devices to find the level plane of a ceiling. These devices include the carpenter's level and laser level.

In the application of acoustical ceilings, a level, as shown in *Figure 20*, is a tool that comes in handy to check the ceiling-grid main runner and cross runner installation. When using a level to install a suspended ceiling, place it at right angles to the runners as you install them. If the tool is perfectly level, the bubble will appear centered between the crosshatches. The leveling should be checked every 6'.

Figure 20 Level.

Many types of laser levels are available that can be used to aid in the installation of suspended ceilings. To use a laser level when installing a ceiling, follow the manufacturer's instructions for the laser level being used. Generally, the procedure involves mounting the laser either on a wall/ceiling mount (*Figure 21*) or on a tall tripod (*Figure 22*). The laser beam is rotated either at the finished ceiling height or at a reference point. A special target is snapped to a grid, and then the grid is moved up or down until the laser beam crosses the target's offset mark. The grid is then secured in place.

Figure 21 Laser level—wall or ceiling mount.
Source: DEWALT Industrial Tool Co.

Figure 22 Laser level—tripod mounted.
Source: vivooo/Shutterstock

WARNING!

OSHA (Occupational Safety and Health Administration) *CFR 1926.54* covers safety regulations for the use of lasers. Some guidelines are as follows:

- Avoid direct eye exposure to the laser beam.
- Only qualified and trained personnel are permitted to operate laser equipment.
- Place a standard laser warning sign conspicuously at major approaches to the instrument use area.
- Always turn off the laser when transmission of the beam is not required.

| 3.4.0 | **Installing Suspended Ceiling Systems** |

Observing the following general guidelines will help to achieve the desired level of professionalism when installing ceilings:

- Ceiling tiles should be arranged so that units less than one-half width do not occur unless otherwise directed by the reflected ceiling plans or job conditions.
- All tiles, tile joints, and exposed suspension systems must be straight and aligned.
- All acoustical ceiling systems must be level to $\frac{1}{8}$" in 12'.
- Tile must be neatly scribed against butting surfaces and to all penetrations or protrusions where moldings are not required.
- Tile surrounding recessed lights and similar openings must be installed with a positive method to prevent movement or displacement of the tiles.
- Tiles must be installed in a uniform manner with neat hairline-fitted joints between adjoining tiles.
- Wall moldings must be firmly secured, the corners neatly mitered, or corner caps used, if preferred.
- The completed ceiling must be clean and in undamaged condition.

Doing overhead work often requires the use of scaffolding or ladders to reach the work area. Working from an elevated platform adds an element of danger to the job. It is important to always place safety first. Use equipment that is the correct height for the job and be sure that all equipment is in good working order. Any defective items must be tagged right away so they will not be used by mistake. Scaffolding must be inspected by a competent person before it is used on each shift, whenever it is moved, and whenever it is changed in any way. Some scaffolding must be designed by an engineer, so know the practices at your worksite. If you are unsure of the site requirements, ask your supervisor or foreman before you begin work.

When working aboveground, always wear the appropriate personal protective equipment. Maintain control of materials and stay alert for personnel working on the ground below you. Resist the urge to over-reach for your equipment. If working on the ground when overhead work is being done, stay alert for falling debris or tools.

NOTE

Make sure all inspections for areas and/or equipment above the grid line have been completed before beginning the ceiling installation.

CAUTION

Be careful when removing ceiling material from its packaging and when handling it. Keep all ceiling material and the grid system clean and undamaged.

WARNING!

Scaffolds must be used and assembled in accordance with all local, state, and federal/OSHA regulations. OSHA requires the building, moving, or dismantling of all scaffolding to be supervised by a competent person who has the training, knowledge, and experience to identify hazards on the jobsite and the authority to eliminate them. Mobile scaffolds, such as baker's scaffolds, should only be used on level, smooth surfaces that are free of obstructions and openings. OSHA regulations also require that mobile scaffold casters have positive locking devices to hold the scaffold in place. When moving a mobile scaffold, apply the moving force as close to the scaffold base as possible to avoid tipping it over. Never move a scaffold when someone is on it.

3.4.1 Exposed Grid Systems

The following sections describe the general procedure for installing an exposed grid ceiling system, also called a direct-hung system.

Install an exposed grid suspended ceiling system according to the following guidelines:

Step 1   Check the room number and location.

- Ensure the correct ceiling is going into the correct room.
- Refer to the construction drawings, reflected ceiling plan, or shop drawing to determine the correct height for the ceiling.
- Check the electrical drawings to ensure the lights will be placed as indicated on the reflected ceiling plan or shop drawing.
- Check the specification sheets or shop orders for special instructions to ensure the proper hangers and fasteners are provided.

Step 2   If needed, install the scaffold.

- Set the scaffold at the correct height to permit the driving of inserts or other fasteners into the overhead ceiling and to permit the connection of hangers.
- Set the lower portion of the scaffold to permit installation of the grid members and ceiling tiles.

Step 3   Establish benchmarks. In some situations, the floor may not be level in all areas of the room you are working in. Make sure to establish benchmarks that are exactly the same height throughout the room and are consistent in all rooms. These benchmarks can be located with the use of a laser level. The benchmarks will provide an accurate starting point for locating the desired height of the finished ceiling.

- Using the level of choice, locate benchmarks at each end of the room near the corners. If the room is extremely large and long, locate the benchmarks at 15' to 20' intervals.
- It is best to locate the benchmarks at eye level (about 5' above the floor).
- Always put benchmarks on each wall and also on protruding walls. It is better to have too many than not enough.

Step 4   Establish the height for the top of the wall angle.

- Once all benchmarks are located and marked, establish a measurement to the bottom of the wall angle.
- To that measurement, add the height of the wall angle.
- Measure from each benchmark and establish a mark at the top of the wall angle. Place a mark at various intervals on the wall.
- Snap a chalk line on those marks. This will establish the top of the wall angle. Be sure to snap the chalk line at the top of the wall angle so that the chalk line will not be visible when the wall angle is installed.

Step 5   Install the wall angle.

- Be sure that the top of the wall angle is even with the chalk line at all points.
- Nail or screw the wall angle to the wall. The wall angle should fit securely to prevent sound leaks.
- Miter the wall angle to fit at the corners or use corner caps to cover the joints.

Install the wall angle according to the following procedure:

Step 1   Square the room by first establishing length and width centerlines on the floor or ceiling, as presented in the Section 3.2.2, *Establishing Room Centerlines*.

NOTE

The procedures given in this module are general in nature and are provided as examples only. Because of the differences in components made by different manufacturers, it is important to always follow the manufacturer's installation instructions for the specific system being used.

NOTE

Omit Steps 3 and 4 if a laser level is being used to establish the correct and level ceiling height, as described in Section 3.3.2, *Ceiling Leveling Equipment*.

NOTE

The following procedure assumes that the centerlines have been laid out on the floor. The procedure would be done in a similar manner if the centerlines were laid out on the ceiling. The only difference is that the centerlines would be transferred down from the ceiling, instead of up from the floor, to the level of the suspended ceiling grid.

Step 2   Transfer the positions of the centerlines from the floor/ceiling to the level required for the suspended ceiling grid as follows:

- At either one of the long walls, use a plumb bob to locate the center of the wall just above the wall angle. Do this by moving the plumb bob until it is directly over the centerline marked on the floor (*Figure 23*). Then, mark the wall just above the wall-angle flange and insert a nail behind the wall angle at this point. Repeat the procedure at the opposite long wall. Run and secure a taut dry line between the nails in the two long walls.
- Repeat the preceding procedure for the two short walls and run a second dry line perpendicular to the first.
- Make a final check using the plumb bob at all four points.

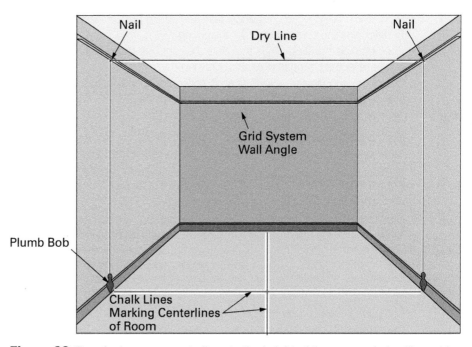

Figure 23   Transferring room centerlines to the height of the suspended ceiling grid.

When completed, you will have two dry lines intersecting at right angles at the center of the ceiling. Use the 3-4-5 method to ensure that the intersection of the dry lines is square. This is important because these intersecting dry lines will be used as the basis for all subsequent measurements used to lay out the ceiling grid system.

In some cases, reflected ceiling plans or shop drawings may indicate a centerline and specify border width. In these instances, follow the suggested layout. If you have neither to follow, you will need to establish the centerline and plan the layout of the ceiling so that the border tiles adjacent to the facing wall are the same width and are not less than one-half the width of a full tile.

"Seeing" through the Ceiling

It is a good idea to keep a package of colored thumbtacks in your toolbox when installing a suspended ceiling. When other trades may have to get back in to complete their work, insert different colored thumbtacks in ceiling tiles to mark the locations of electrical and mechanical services. The marking will allow the other trades to locate their components without having to raise a lot of tiles and will therefore help minimize finger smudges and damage to the tiles.

One method for determining the width of the border tiles is to convert the wall measurement from feet to inches, then divide this amount by the width of the ceiling tiles. For example, assume the room measurements are 42'-6" × 30'-6".

What will be the width of the border tiles running parallel to the two 42'-6" walls if 2' × 2' tiles are being used? To find the answer, proceed as follows:

Step 1 Take the measurement of the short wall (30'-6") and convert feet to inches:

$$30' \times 12" = 360" + 6" = 366"$$

Step 2 Divide that amount by 24" (the width of a tile):

$$366" \div 24" = 15 \text{ tiles with a remainder of } 6"$$

Step 3 If the division does not result in a whole number, add the width (in inches) of a full board to the remainder (in this case, it is 6"):

$$6" + 24" = 30"$$

Step 4 Divide this by 2; the result is 15. This would be the width of a border tile on each side. When adding the width of a tile to the remainder in Step 3, you must delete one full tile from the total. This means that there would be 14 full tiles plus a 15" border on each end (*Figure 24*):

$$14 \times 2' + 30" = 30'\text{-}6"$$

Step 5 Determine the width of the border tiles for the other two remaining walls in the same manner.

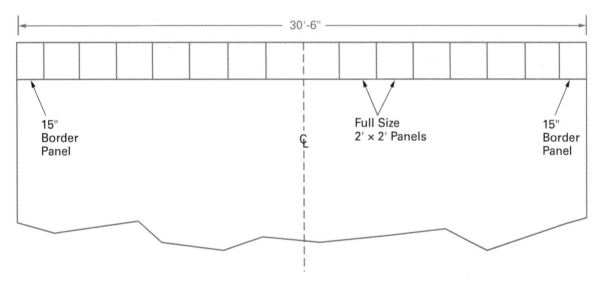

Figure 24 Fitting the tiles to the room.

Once all border unit measurements have been established, it is necessary to install eye pins, also known as ceiling clips or hanger clips, to secure the hangers. Prior to installing the eye pins, you must locate the position of the main runners and cross runners. When the direction of the main tees and cross tees is not indicated on the plans, the installer will decide the best direction. The stronger main runners are more expensive, so installers usually choose to run the main tees in the shortest direction to minimize use of materials. In *Figure 25*, the plans indicated the main runner should run in the long direction. Line A in *Figure 25* represents the main runner, and Line B represents the first cross runner line.

Step 1 Locate the first dry line (A) as shown in *Figure 25*. The dry line should be installed approximately 1⅛" above the flange of the wall angle. The dry line will also be used to indicate the bend in the hanger wire.

Step 2 Fasten a second dry line (B) so that it intersects line A at right angles with the proper measurement for the border tiles.

Step 3 The intersection point will be the location of the first hanger wire. Set the first eye pin directly over the intersection of dry lines A and B. Install the other eye pins every 4' on center. Be sure to follow dry line A as a guide for the locations.

NOTE

Most main runners measure 1⅛" from the bottom of the runner to the hole for the hanger wire.

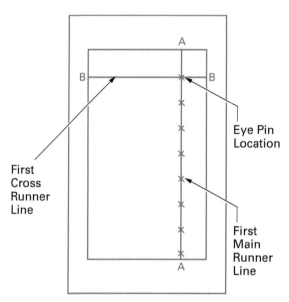

Figure 25 Locating the positions of the first main runner and cross runner.

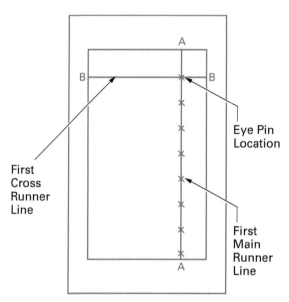

WARNING!

Powder-actuated tools (PATs) are to be used only by trained operators in accordance with the operator's manual. Operators must take the following precautions to protect both themselves and others when using powder-actuated tools:

- Operate the tool as directed by the manufacturer's instructions and use it only for the fastening jobs for which it was designed.
- To prevent injury or death, make sure that the drive pin cannot penetrate completely through the material into which it is being driven.
- To prevent a ricochet hazard, make sure the recommended shield is in place on the nose of the tool.

NOTE

At least three turns should be made when twisting the wire. The wire should also be tight to the runner.

NOTE

As indicated previously, in most exposed grid systems, the main runners are available with pre-punched holes. The upper edges of the holes generally measure 1⅛" from the bottom of the runner; the dry line will have to be adjusted prior to bending the hanger wire.

As you install each eye pin, suspend the hanger wire from the eye pin. Use pretied wire whenever possible. If not using pretied wire, as you install each eye pin and wire, allow enough wire to ensure proper tying to the eye pin and the main runner. These two ties will require from 10" to 24" of wire. More hanger wire may be required if the ceiling is hung in part from structural members along with the eye pins. All hanger wire is precut and can be obtained in various lengths and gauges. Refer to the specifications for the gauge of hanger wire to use. There are cases in which there may be cast-in-place inserts for hanger wires. In such cases, the insertion of eye pins and installation of wire will not be necessary.

The eye pins and hanger wires should be located on 4' centers in both directions. In most cases, it is best to hang both at the same time, as it requires less movement of the scaffold. To mark and bend the wires, proceed as follows:

Step 1   Mark the hangers where they touch the dry line.

Step 2   Twist the wire using side cutters and bend the other end to the mark (*Figure 26*).

Install the main runners as follows:

Step 1   Measure and cut the main runner so that the cross runner will be at the proper distance for the border tile.

Step 2   Suspend the first length of main runner from the first row of hangers. It is important that the first few hanger wires be perpendicular to the main runner. If they are not, use a Whitney punch and make new holes.

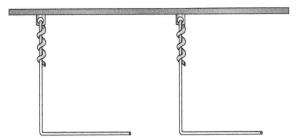

Figure 26 Bending hanger wire.

Step 3   Continue to install the balance of the first row of main runners; splices will be needed. Note that various types of splices are used with various grid systems and are supplied by the grid system manufacturer.

Step 4   After installing the last full length of main runner, measure, cut, and install an end piece from a full length of main runner to complete the first run.

Step 5   After each end of the main runner is resting on the wall angle, insert all hanger wires and twist the wires to secure them in place (*Figure 27*).

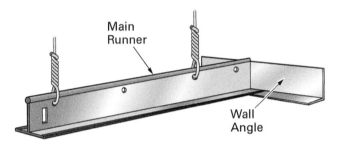

Figure 27 Inserting hanger wires in the main runner.

Use the following procedure to install the cross runners:

Step 1   Measure the distance from the main runner to the wall angle. This width was determined earlier when completing the room layout.

Step 2   Using the snips, cut the cross runner to the correct border width. Be sure to cut and save the correct end or it will not slide into the main runner.

Step 3   Insert the factory end into the main runner and let the other end rest on the wall angle (*Figure 28*).

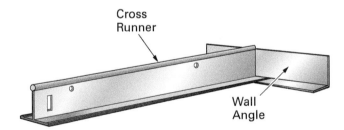

Figure 28 End of cross runner resting on wall angle.

Step 4   When the main runner and the cross runner are perpendicular to each other, lock the cross runner in place. Install the remaining cross runners between the main runners.

Step 5   To stabilize the grid system, install pop rivets at every other cross runner.

Step 6   Double-check the squareness of the entire grid.

Hold-Down Clips

Some ceiling tiles require hold-down clips to secure the ceiling tiles to the grid. For example, clips are used with lightweight tiles to prevent them from reacting to drafts. One manufacturer specifies that clips be used if the tiles weigh less than 1 pound. Hold-down clips are not necessarily required for ceilings used in fire-resistance-rated applications. Check the manufacturer's instructions.

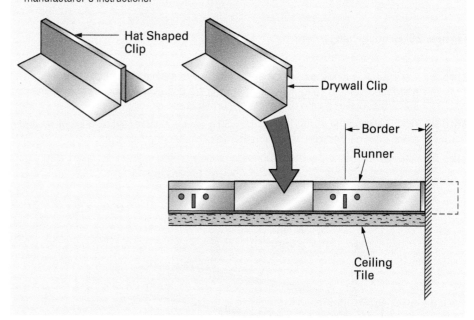

Did You Know

Squaring It Up

If the main runners and cross runners in a ceiling grid are out of square, the tiles will not fit properly into the grid, which may result in gaps between the grid and the tiles. Some quick measurements and a little math can be used to prevent this problem.

You have likely heard of the Pythagorean theorem:

$$a^2 + b^2 = c^2$$

The letters in the formula each represent one side of a triangle. The letters a and b are the straight sides and the letter c is the diagonal side. In a ceiling grid, the measurements of side a and side b would be sections of cross runners and main runners. Square (multiply them by themselves in mathematical terms) these two measurements and add them together. Measure the diagonal side of the imaginary triangle, side c, and square this measurement. If that section of the ceiling grid is perfectly perpendicular (squared in construction terms), the square of side c should equal the added squares of sides a and b. If these measurements are not equal, make slight adjustments to the main runners and cross runners to achieve squareness.

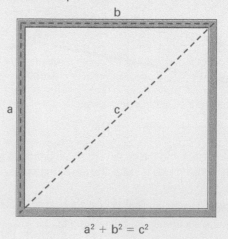

$$a^2 + b^2 = c^2$$

NOTE

Some grid systems use automatic locking devices to secure the cross runners into the main runners. Other grid systems use clips. Once all cross runners have been installed into the main runners, the grid system is ready to receive the ceiling panel or tile.

Prior to installing the ceiling tiles, wash your hands to keep the tiles clean. It is good practice to wear white gloves when handling ceiling tiles. Start by cutting the border tiles and inserting them into position. After the border tiles are installed, proceed to install the full tiles. As the full tiles are being placed, install hold-down clips, if required. General installation guidelines include the following:

- Install the ceiling tiles in place according to the job progression. Do not jump or scatter tiles in the grid system.
- Exercise care in installing tiles so as not to damage or mar the surface.
- Always handle tiles at the edges and keep your thumbs from touching the finished side of the tiles. If gloves are not worn, use cornstarch, powder, or white chalk on your hands.
- If the ceiling tile has a deeply textured pattern, insert it into the grid system so that the directional pattern will flow in the same direction. Such matching of ceiling tiles adds beauty to the finished suspended ceiling.

Upon completion of the ceiling, clean up the work area as follows:

Step 1   Dismantle the scaffold.

Step 2   Pick up all tools and equipment.

Step 3   Secure all equipment in a safe place.

Step 4   Pick up all trash.

Step 5   Sweep up any dust or debris.

3.4.2 Metal-Pan Systems

Metal-pan systems use $1^1/_2$" U-shaped, cold-rolled steel furring channel members for pan support. They are normally installed 4' on center. However, it may be necessary to install them at lesser distances depending on the location of the light fixtures. The furring channels are installed by looping the hanger wires around them. Twist and secure using saddle wire. It is important that the furring channels are properly leveled to ensure a level ceiling.

Once the furring channels have been installed, the main runners (tee bars) must be placed. These bars are manufactured, cold-rolled or zinc-coated steel or aluminum. They are fastened to the furring channels using special clips. The tee bars run at right angles to the furring channels (*Figure 29*).

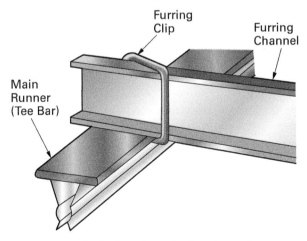

Figure 29 Main runner (tee bar) clipped to a furring channel.

The tee bar has a spring-locking feature, which grips the metal pan. This feature allows the pan to be removed for access to the area above the pan. If the room plan dimensions are longer than the length of the tee bar, use a tee bar splice to couple the tee bars together, extending them to the required length (*Figure 30*).

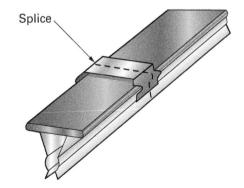

Figure 30 Tee bar splice.

High-Durability Ceilings

Highly durable ceiling tiles provide for long life and easy maintenance wherever ceilings are subjected to improper use, vandalism, or frequent removal for plenum access. They are used in applications where resistance to impact, scratches, and soil are major considerations. Ceilings in areas such as school corridors or gymnasiums need to withstand abuse, including surface impact. In any areas where lay-in ceiling tiles frequently need to be removed for plenum access, scratch-resistant tiles are highly desirable. Otherwise, the tile surface can be scratched, scuffed, or chipped as it is slid across the metal suspension system components. Ceilings installed in laboratories, clean rooms, and food preparation areas are normally required to meet special standards to ensure they can withstand repeated cleaning.

Use the following procedure to install the tee bars:

Step 1 Install the first tee bar at right angles to the $1\frac{1}{2}$" furring channel. Be sure it is correctly aligned.

Step 2 When in position, place tee bar clips over the furring channel and insert the ends under the flange of the tee bar. Hang additional clips in the same manner along the length of the bar.

Step 3 Install the second tee bar parallel to the first one. The spacing should be from 1' to 4', depending on the size of the metal pans.

After installing all the tee bars in one section of the room (depending on the working area of your scaffold), begin installing the metal pans. Take care in handling the pans. Use white gloves or rub your hands with cornstarch to prevent any perspiration or grease marks from marring the surface of the pans. If care is not taken, fingerprints will be plainly visible when the units are installed. Use the following procedure to remove the pans from their container:

Step 1 Place the pan's finished surface face down on some type of raised platform, such as a table, which has been covered with a pad to protect the surface of the pan.

Step 2 Insert the wire grid into the back side of the metal pan. The wire grid is installed between the metal pan and backing pad to provide an air cushion between the two surfaces.

Step 3 Over the grid, place the paper- or vinyl-wrapped mineral wool or fiberglass batt or pad.

When installing the metal pans, begin at the perimeter of the ceiling, next to the walls, because the units must fit into the channel wall angle. Install the pans as follows:

Step 1 From the room layout or reflected ceiling plan, obtain the width of the border units. Measure the pan to this width.

Step 2 Using a band saw, cut the pan $\frac{1}{8}$" short of the desired width along the edge to be inserted into the wall angle.

Step 3 Slide the pan into the wall angle. Do not force the pan all the way into the molding. Leave $\frac{1}{8}$" for expansion.

Step 4 Insert two crimped edges of the metal pan into the spring-locking tee bars.

Step 5 When the pan is in position, insert the spacer clip into the channel-molding cavity and over the cut edge of the pan. This will prevent the pan from buckling along this cut edge.

Step 6 Cut the wire grid and backing pad to fit the border unit. Place the unit into position.

Step 7 At the corners of the room, install the pans in the order shown in *Figure 31*.

NOTE

In some cases, the metal pans come with the wire grid and pad already assembled.

Indoor Swimming Pools

All-aluminum grid systems are not recommended for use above indoor swimming pools because chlorine gases cause aluminum to corrode.

1	3
2	4

Figure 31 Tile layout.

Once the perimeter pans are installed in each row, the full-size pans can be put into place as follows:

Step 1   Grasping a pan at its edges, force its crimped edges into the tee-bar slots. Use the palms of your hands to seat the pan.

Step 2   After installing several of the pans as noted in Step 1, slide them along the tee bars into their final position. Use the side of your closed fist to bump the pan into level position if it does not seat readily.

Step 3   If metal pan hoods are required, slip them into position over the pans as they are installed. The purpose of the hood is to reduce the travel of sound through the ceiling into the room.

If a metal pan must be removed, a pan-removal tool is available (*Figure 32*). To pull out a pan, insert the free ends of the device into two of the perforations at one corner of the pan and pull down sharply. Repeat this at each corner of the pan. Following this removal procedure eliminates the danger of bending the pan out of shape. *Figure 33* shows a finished metal-pan ceiling.

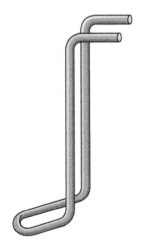

Figure 32 Pan-removal tool.

Figure 33 Metal-pan ceiling.
Source: David R./Alamy

Placing Main Runners

One way to locate the position of the first main runner is to convert the width of the room to inches and divide it by 48 (assuming you are installing 48" tiles). Then add 48" to any remainder and divide that result by 2 to obtain the distance from the wall to the first main runner. The rest of the main runners are then placed at 4' intervals. Try an example assuming the width of the room is 22'.

Converting 22' into inches yields 264". Dividing that result by 48 gives 5' with a remainder of 6". Adding 48" yields 54", which is then divided by 2. The first main runner is placed 27" from the wall.

3.4.3 Direct-Hung Concealed Grid Systems

The installation of the concealed grid system begins in the same way as the conventional exposed grid system previously discussed.

Step 1 Once chalk lines have been established on the walls at the required height above the floor, the next step is to install the wall angle. This molding provides support for the grid and panel or tile at the wall, as it does for the other ceiling systems. The molding should be fastened with nails, screws, or masonry anchors. At corners, miter the inside corner and use a cap to cover the outside corner.

Step 2 Lay out the grid and install the hanger inserts and hangers according to the reflected ceiling plan or lighting layout. Once this has been completed, install the concealed main runners as is done for the exposed grid system (see Section 1.1.1, *Exposed Grid Systems*). The main runners are the primary support members. The method of coupling them to attain a specific length will vary with the manufacturer of the system, but they can all be spliced to the desired length.

Step 3 Install the cross-stabilizer bars and concealed cross tees at right angles to the main runners. They rest on the flange of the main tee runners.

Step 4 Place the ceiling panel into position. Cut in the first row of panel at a line perpendicular to the main runners, as previously established. Attach the panel to the wall using spring clips.

Flat splines (*Figure 34*) are metal or fiber units that are inserted by the installer into the unfilled panel kerfs between the concealed main runners. Splines are used to prevent dust from seeping through the kerfs in adjacent panels.

If access is needed to the area above the ceiling, special systems are available that can be incorporated into the ceiling (*Figure 35*).

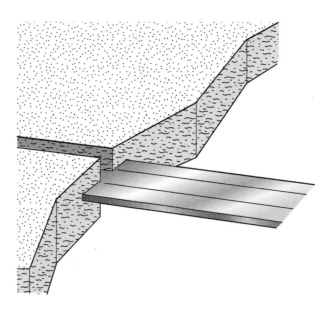

Figure 34 Ceiling-panel spline.

Insulation

Insulation is usually not recommended over mineral ceiling tiles because the additional weight could cause the tiles to sag. If the use of insulation cannot be avoided because of occupancy codes, limit the insulation to R-19 (0.26 pounds per square foot). Use only roll insulation and lay it perpendicular to the cross tees, so that the grid supports the weight of the insulation. If batts are required by code, 24" × 24" ceiling tiles should be used. Check codes carefully before applying insulation in a fire-resistance-rated system.

Facts about Ceiling Tiles

- Ceiling tiles are often called pads.
- Certain lighting fixtures are manufactured in the same sizes as tiles to drop directly into suspended ceilings.
- Some tiles are treated to inhibit the growth of mold, mildew, fungi, and certain bacteria.
- Special tiles are designed for use below sprinkler systems. These tiles will shrink and fall out of the grid as the room temperature rises, allowing the water to reach the fire.
- Most ceiling tiles can be recycled.

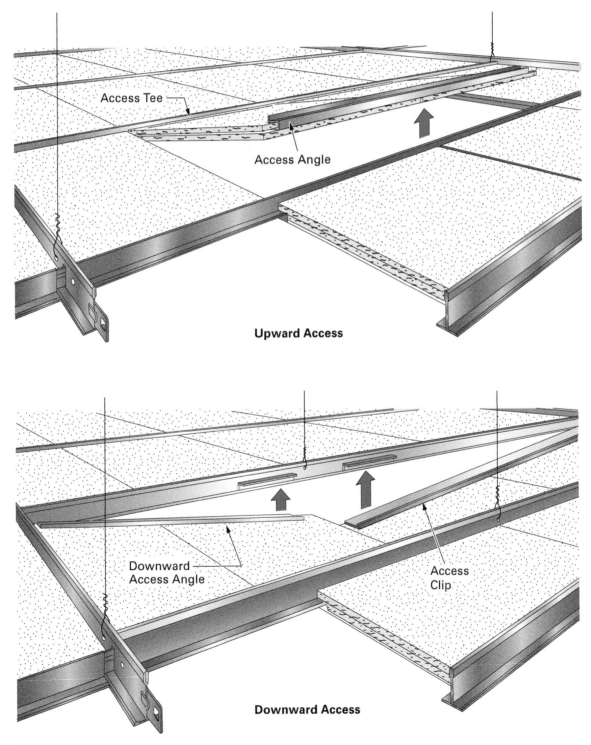

Upward Access

Downward Access

Figure 35 Access for concealed grid ceilings.

3.4.4 Luminous Ceiling Systems

A standard luminous system is installed in the same way as an exposed grid suspension system except for the border cuts. Luminous ceilings are placed into the grid members in full modules. Any remaining modules are filled in with acoustical material that has been cut to size.

Luminous panels are used with a 2' × 2' or 2' × 4' standard exposed grid to provide the light diffusing element in the system (*Figure 36*). These panels are laid in between the runners. Many panel sizes and shapes are available.

Figure 36 Luminous office ceiling.
Source: August0802/Getty Images

A nonstandard luminous ceiling is installed in the same way as a standard system with regard to room layout, hanger insert, installation of hanger wire, main supports, secondary supports, and attachment of some type of wall angle. Because nonstandard ceilings can differ so much in terms of size, complexity of the system, and the exactness of the installation, shop drawings are required. As always, when installing a ceiling system, consult the manufacturer's information for specific details.

3.4.5 Suspended Drywall Ceiling Systems

When drywall is used instead of panels or tiles to cover the surface of a ceiling, it is either fastened directly to the ceiling deck or a drywall suspension grid is installed. Drywall suspension systems are advantageous because they are quick and easy to install in comparison to other ceiling systems, and they can be shaped into the framework for more intricate designs, such as coffered, step soffit (*Figure 37*), and tray ceilings. Suspended drywall ceiling can use either the drywall furring system or the drywall grid system.

Figure 37 Step soffit ceiling.
Source: Dennis Axer/Alamy

A suspended drywall furring system is used when it is desirable or specified to have a drywall finish or drywall backing for an acoustical tile ceiling. *Figure 38* is an example of a suspended drywall furring system. The installation of a suspended furring system requires the use of special carrying channels.

Figure 38 Suspended drywall furring system.
Source: Don Wheeler

Installing Carrying Channels

Carrying channels, also known as $1\frac{1}{2}$" cold-rolled channel or black iron, are used for drywall furring systems to compensate for the additional weight of the systems. To install carrying channels, proceed as follows:

Step 1   Use a laser level to establish the reference line.

Step 2   Install the wall angle, if used.

Step 3   Fasten the hanger inserts into the ceiling structure above. Install a row of hanger inserts 4' on center, or as specified by the manufacturer, into the ceiling structure.

Step 4   Install the hangers. Use No. 8 gauge wire or as specified. Hang the wires from the hanger inserts, then twist and secure them. Allow enough wire to tie the hanger to the insert and the carrying channel. In general, 24" to 28" of wire is adequate.

Step 5   Mark the places where the laser target touches the hanger wires. Mark each hanger wire to the spot where it is to be bent. Bend the wire to a 90° angle.

Step 6   When all the hanger wires in one row have been marked and bent, fasten the carrying channel to the hanger wires (*Figure 39*). Loop the wires around the carrying channel at the point where the wires have been bent. Twist and fasten the wires using a saddle tie. It is important that the carrying channels are installed level.

Step 7   If it is necessary to extend the length of a carrying channel because of the room size, face the open U of the second channel toward the U of the one already installed. Overlap by 8" to 12" and tie together with wire (*Figure 40*).

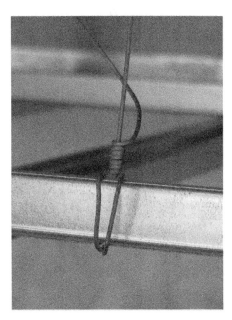

Figure 39 Carrying channel supported by hanger wire.
Source: Don Wheeler

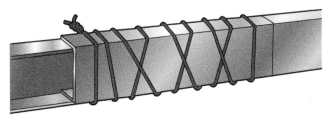

Figure 40 Splicing carrying channels.

Step 8   Continue to install the hanger inserts, wires, and carrying channels on 4' centers over the entire ceiling.

Installing Furring Members and Drywall

After the carrying-channel support members are installed, proceed with the installation of the cross-furring members. There are different kinds of furring members that can be used at right angles to the carrying channels at 16" on center (maximum). For example, a hat furring channel (*Figure 41*) is used when it is desirable to screw the drywall board to the furring channels with self-tapping screws. Install the furring channels as follows:

Step 1 Install the furring channels perpendicular to the carrying channels. Attach the furring channel to the carrying channel using tie wire (*Figure 42*).

Step 2 Continue to install the furring channels 16" on center. If it is necessary to extend the furring channels beyond their normal length, lap one end of the channel over the next piece a minimum of 8" and wrap the splice well with tie wire (*Figure 43*). Avoid installing furring channels over positions where any fixtures are to penetrate the ceiling.

Step 3 Screw the drywall to the furring channels (*Figure 44*). Be sure to use screws of sufficient length. The screw length should be long enough to go through the board and extend through the furring channel about $\frac{5}{8}$" with three exposed threads.

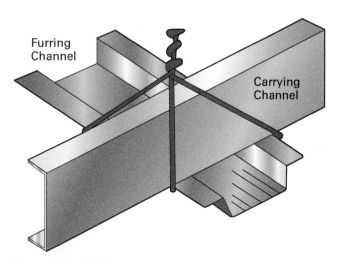

Figure 41 Hat furring channel.

Figure 42 Furring channel attached to carrying channel with clips.
Source: Don Wheeler

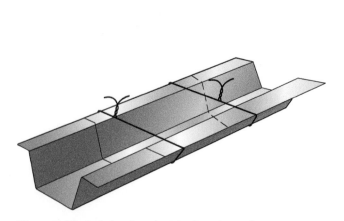

Figure 43 Splicing (lapping) furring channels.

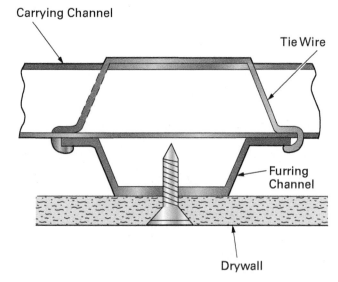

Figure 44 Drywall screwed to a furring channel.

NOTE

Keep acoustic panels clean during installation.

3.4.6 Installing Acoustical Panel

If acoustical panel is to be used as the finish over the drywall, proceed as follows:

Step 1 Lay out the room. Measure the length of one of the two shorter walls. Divide this measurement in half. Mark this halfway point on the drywall next to this point on the wall. Next, measure the length of the facing wall at the opposite end of the room. Divide this wall length in half and mark this point on the drywall.

Step 2 Set up a laser level and align the horizontal beam with the two marks you made on the drywall.

Step 3 Align the vertical laser beam at the midpoint of the horizontal line. If your laser level does not generate two beams simultaneously, use tape to mark the horizontal beam's location and then align a vertical beam to the tape line.

Step 4 Install the ceiling panel per the manufacturer's instructions beginning at the junction of the two centerlines.

3.4.7 Installing Furring Channels Directly to Structural Members

Sometimes furring channels are installed directly to structural members such as open-web steel joists or wood I-joists instead of suspended carrying channels. In these instances, install the furring channels as follows:

Step 1 Mark the walls to indicate the height to be used to level the ceiling. Set the ceiling-height control marks to indicate the position of the bottom edge of the furring channels. Check the level of the structural joists. If they are not level, use the low point as the common level for installing the furring channels.

Step 2 Measure up from the benchmarks to the height of the lowest point on the joists. Determine the distance of the low joist point minus the height of the furring channel and mark it on all walls to indicate a uniform level for the lower edge of all furring channels. Be sure to measure up from the benchmarks to establish this common level on all walls.

Step 3 Establish the level lines using the ceiling control marks.

Step 4 Install the wall angles (if used). New level lines may have to be used to guide the installation process.

Step 5 Install the furring channels in the same manner as described in *Installing Furring Members and Drywall* in Section 3.4.5, with the following exceptions:

- When attaching furring channels to steel joists, tie them together with the appropriate-gauge tie wire. Wrap the wire around the two so that it bridges the joist and supports the furring channel on either side of the joists. Tie the wire ends together at the side of the union. Furring channel can also be attached to steel joists using an 8" overlap and four #8 screws (two screws through each of the small $1/2$" flanges).
- Shimming may be needed between the joists and the furring channel. Be sure to check the level of the furring channel along its entire length. Shim where needed to correct any deviation from level.

3.4.8 Drywall Suspension Systems

The drywall grid system (DGS) is a suspension ceiling system for drywall ceilings. In this system, a grid is suspended from the ceiling, and the drywall panels are installed on the grid (*Figure 45*). Suspending the drywall below the ceiling deck, as opposed to attaching it directly, allows ductwork and other items to remain in place while still achieving the finished drywall look.

This system is ideal for large ceilings, long flat ceilings, and floating ceilings. It can also be used in small areas, called short-span areas, such as hallways. In addition, DGS can be used to create dramatic, curved ceilings and distinctive designs.

Figure 45 Suspended ceiling drywall grid system (DGS).
Sources: (left) Dennis Axer/Alamy; (right) Sever180/Shutterstock

Special Furring Systems

Some manufacturers make furring-system cross tees in 14", 26", and 50" lengths. These sizes can reduce the time it takes to install an F-type lighting fixture from as long as 30 minutes to less than 1 minute.

Drywall grid systems are installed much like the standard exposed grid system and use the same size main tees, cross tees, and wall angle. However, the components for this system are more particular and are normally purchased from specialized suppliers. As with all proprietary systems, consult the approved construction documents and manufacturer's information for specific installation instructions.

An extruded aluminum drywall perimeter trim used in conjunction with the DGS provides a great finished edge for floating ceilings with straight or curved edges. Curved perimeter trim is factory pre-curved to the desired radius/radii. This trim will not only create a solution at the perimeter edge but is also a labor cost reducer and time saver that will ensure a better-looking perimeter edge than drywall. The perimeter trim is secured to the drywall suspension with factory clips, which are screwed into the drywall for a seamless appearance. Clean and crisp perimeter edges are a guarantee if extruded trims are used. These trims are generally factory primed and field-finished along with the drywall.

3.4.9 Ceiling Cleaning

Immediately after installation and after extended periods of use, suspended ceilings may require cleaning to maintain their performance characteristics and attractiveness. Dust and loose dirt can easily be removed by brushing or using a vacuum cleaner. Vacuum cleaner attachments such as those designed for cleaning upholstery or walls do the best job. Make sure to clean in one direction only to prevent rubbing the dust or dirt into the surface of the ceiling.

After loose dirt has been removed, pencil marks, smudges, or clinging dirt may often be removed using an ordinary art gum eraser. Most mineral-fiber ceilings can be cleaned with a moist cloth or sponge. The sponge should contain as little water as possible. After washing, the soapy film should be wiped off with a cloth or sponge slightly dampened in clean water. Vinyl-faced fiberglass ceilings and Mylar™-faced ceilings can be cleaned with mild detergents or germicidal cleaners.

Cleaning Ceiling Tiles

The materials and methods used for cleaning ceiling tiles vary widely, depending on the finish and texture of the tiles. One manufacturer prescribes eight different cleaning methods for its line of ceiling tiles. It is very important to check the manufacturer's instructions before proceeding. When cleaning grids, the ceiling tiles should be removed to prevent cleaning solution or dirt from getting on the tiles. Always wear gloves when handling or cleaning tiles.

3.0.0 Section Review

1. Which soil classification is considered the most stable?
   a. Class A
   b. Class B
   c. Class D
   d. Class F

2. If a room does not have a flat ceiling, the centerlines can be laid out on the floor and then transferred upward using _____.
   a. a laser level
   b. a plumb bob and dry lines
   c. a story pole
   d. triangulation (the 3-4-5 rule)

3. When installing the grid for a suspended ceiling, you should use a level to check the installation at intervals of _____.
   a. 3'
   b. 4'
   c. 6'
   d. 8'

4. After the first eye pin, where should other eye pins be installed?
   a. 2' on center
   b. 4' on center
   c. 6' on center
   d. 8' on center

Module 45203 Review Questions

1. From their source, sound waves travel _____.
   a. on a line of sight
   b. in all directions
   c. vertically
   d. at right angles

2. The intensity of a sound wave refers to _____.
   a. the number of wave cycles it completes in a second
   b. its loudness
   c. its frequency
   d. the degree of loudness or softness

3. A sound level of 100 dBA is considered _____.
   a. very loud
   b. moderately loud
   c. loud
   d. quiet

4. The term used by manufacturers to compare the noise absorbency of their materials is the _____.
   a. ceiling attenuation class (CAC)
   b. articulation class (AC)
   c. noise reduction coefficient (NRC)
   d. sound transmission classification (STC)

5. A(n) _____ is used if the grid system is to be concealed from view.
   a. metal-pan system
   b. direct-hung ceiling system
   c. integrated ceiling system
   d. luminous ceiling system

6. All ceilings, walls, pipe, and ductwork in the space above a luminous ceiling will normally be painted with a _____.
   a. light-refracting finish
   b. 75% to 90% reflective matte white finish
   c. 90% to 100% reflective matte finish
   d. luminous matte finish

7. _____ are inserted into the main runners at right angles and spaced an equal distance from each other, forming a complete grid system.
   a. Hangers
   b. Wall angles
   c. Hold-down clips
   d. Cross runners

8. In an exposed grid ceiling system, hangers are used to support the _____.
   a. 4' cross tees
   b. 2' cross tees
   c. wall angle
   d. main runners

9. Acoustical panels and tiles used in suspended ceilings stop the transmission of unwanted sounds by the process of _____.
   a. reflection
   b. reverberation
   c. absorption
   d. refraction

10. Butt-jointed tiles that produce a monolithic ceiling design with no visible support system are known as _____.
    a. beveled edge tiles
    b. concealed tee tiles
    c. lay-in tiles
    d. profiled edge tiles

11. Which of the following is a safety guideline for using a baker's scaffold?
    a. Never ride on a moving scaffold.
    b. Inspect the scaffold once per week.
    c. Apply pushing and pulling force at or near the top of the scaffold.
    d. Erect a ladder on top of the scaffold to increase your reach or height.

12. A(n) _____ shows the features of the ceiling while keeping those features in proper relation to the floor plan.
    a. elevation plan
    b. mechanical plan
    c. reflected ceiling plan
    d. orthographic ceiling plan

13. If job conditions do not agree with what is shown in the specifications, you should _____.
    a. change the job conditions to match the specifications
    b. match the specifications as closely as you can
    c. change the specifications to match the job conditions
    d. contact your supervisor and ask for instructions on how to proceed

14. The points at which electrical and plumbing risers penetrate a ceiling are shown on _____.
    a. MEP drawings
    b. floor plans
    c. section views
    d. elevation drawings

15. For a suspended ceiling that is 10' × 16', using ceiling tiles that are 2' × 4', with the main runners parallel with the ceiling's long dimension and the 4' length of the tiles parallel with the ceiling's short dimension, the first main runner should be installed at a distance from the wall of _____.

    a. 18"
    b. 24"
    c. 36"
    d. 48"

16. For a suspended ceiling that is 10' × 16', using ceiling tiles that are 2' × 4', with the main runners parallel with the ceiling's long dimension and the 4' length of the tiles parallel with the ceiling's short dimension, the dimensions (width and length) of the border ceiling tiles to be installed along the long dimension of the ceiling will be _____.

    a. 18" × 24"
    b. 24" × 30"
    c. 24" × 36"
    d. 36" × 48"

17. When correctly laid out, the centerlines for a grid system will intersect at a _____ at the exact midpoint of the room.

    a. 25-degree angle
    b. 45-degree angle
    c. 90-degree angle
    d. 120-degree angle

18. When laying out eye pins for hanger inserts, two dry lines are used to determine the _____.

    a. exact center of the room
    b. positions of all cross ties
    c. centerline of the first ceiling tile
    d. location of the first wall angle

19. When installing the metal pans in a metal-pan ceiling system, begin by installing them _____.

    a. in the middle of the ceiling
    b. at the perimeter of the ceiling next to the walls
    c. at the corners of the room
    d. at the most convenient location

20. Splines installed into the unfilled panel kerfs in a direct-hung concealed grid ceiling system _____.

    a. prevent buckling of the ceiling
    b. are used in place of stabilizer bars
    c. prevent shifting of the panels
    d. prevent dust from seeping through the kerfs in adjacent panels

Answers to odd-numbered review questions are found in the Module Review Answer Key found at the end of this book.

Answers to Section Review Questions

Answer	Section	Objective
Section One		
1. c	1.1.4	1a
2. a	1.2.0	1b
3. b	1.3.1	1c
Section Two		
1. a	2.1.0	2a
2. d	2.2.0	2b
Section Three		
1. a	3.1.0	3a
2. b	3.2.2	3b
3. c	3.3.2	3c
4. b	3.4.1	3d

Interior Specialties

Source: Photos provided courtesy of Armstrong World Industries

Objectives

Successful completion of this module prepares you to do the following:

1. Describe the composition and use of common types of specialty interior products as well as their care and installation methods.
   a. Explain the acoustical benefits of fabric-covered fiberglass wall panels.
   b. Understand why veneer is often used instead of solid wood as an interior treatment.
   c. Explain why glass fiber reinforced gypsum is a popular choice for ornate finishes.
   d. Describe what gives Tectum® its durability and humidity resistance.
   e. Understand the applications for fabric wall and ceiling treatments.
   f. Explain how metal column covers, reveal trim, and CNC routers are used to enhance the aesthetics of a space.

2. Describe the specialty ceiling systems covered in this module, including their care and installation methods.
   a. Explain the visual benefits of using a trim system on a suspended ceiling.
   b. Describe the suspension system of a linear ceiling.
   c. Understand that certain software programs allow contractors to easily design custom dome and vaulted ceilings.

3. Describe additional safety hazards associated with interior specialties.
   a. Identify key hazards associated with installing different types of interior specialties.

Performance Tasks

Under supervision, you should be able to do the following:

1. Make (or describe in writing the steps involved to make) a simple vinyl- or fabric-covered interior panel, including a hole to accommodate a fixture.

2. Describe in writing the parts and elements of a linear ceiling system.

3. Review the safety data sheet (SDS) of a common adhesive or solvent used for interior finish systems. Identify the hazards of the material and mitigation measures (e.g., using a ventilator).

Overview

The number and variety of interior finish products is almost endless. In commercial buildings, finish products need to be attractive and easy to install and maintain. They should support the desired acoustical qualities of an area. Most finish products used in commercial buildings incorporate some type of design or material that improves an area's acoustical properties. This module provides information about some of the most frequently used wall and ceiling systems, including fabric-covered panels, wall upholstery, and wood panels and slats. General installation procedures are also covered. Unique finish products, such as ceiling domes and architectural features, are briefly discussed.

Digital Resources for Drywall

Scan this code using the camera on your phone or mobile device to view the digital resources related to this craft.

1.0.0 Interior Finish Systems

Performance Task

1. Make (or describe in writing the steps involved to make) a simple vinyl- or fabric-covered interior panel, including a hole to accommodate a fixture.

Objective

Describe the composition and use of common types of specialty interior products as well as their care and installation methods.

a. Explain the acoustical benefits of fabric-covered fiberglass wall panels.

b. Understand why veneer is often used instead of solid wood as an interior treatment.

c. Explain why glass fiber reinforced gypsum is a popular choice for ornate finishes.

d. Describe what gives Tectum® its durability and humidity resistance.

e. Understand the applications for fabric wall and ceiling treatments.

f. Explain how metal column covers, reveal trim, and CNC routers are used to enhance the aesthetics of a space.

Interior finish work makes commercial areas attractive and more usable. The number of interior finish products available is staggering. Although looks are important, the most important characteristic of any large space is its acoustics. In the past, a great deal of time and effort went into interior finish work, and much of it was done by hand. Today, an interior finish system must make an area attractive and improve its acoustics, but it also should be easy to install.

Manufacturers spend considerable time and money designing systems that are easy to install and maintain while looking attractive and unique. Systems can often be used as both ceilings and wall coverings. Keep the following points in mind when installing any finish system:

- Neatness counts. Before working with any system, sweep or vacuum the installation area.

- Protect the finish system from dirt, scratches, and other damage.

- Handle materials with care to avoid damaging their acoustic properties and finish.

- Store all finish materials in a clean, safe location.

- Wash your hands before handling any easily soiled materials.

1.1.0 Vinyl- and Fabric-Covered Fiberglass Panels

Fiberglass panels (*Figure 1*) are mounted on walls and ceilings to change the acoustical properties of a room. These panels are often used in large, open areas that would normally reflect and echo sounds. Areas such as auditoriums and churches, or places that are often crowded with people in conversation, such as reception rooms, can often benefit from these panels. Panels are available in many sizes and shapes and are covered with fabric or vinyl to complement the decorating scheme of an area.

Figure 1 Fabric-covered fiberglass wall panels.
Source: Tavarius/Shutterstock

1.1.1 Care

Fabric-covered wall panels can be easily damaged, so be careful when handling them and follow these basic guidelines:

- Store panels in an inside area on the jobsite where they will be protected from moisture and humidity.
- Always store panels flat and covered to protect the fabric from dust and dirt.
- Carry panels upright and by their edges to avoid bending them during installation. Use a two-person carry or a drywall cart, if necessary.
- Protect the fabric or vinyl during installation. Wash your hands and wear clean white gloves to keep the panels unsoiled. Do not install damaged panels.

1.1.2 Surface Preparation

Before installing the panels, scrape any loose or peeling paint from the wall and ceiling surfaces, and clean off any dust, dirt, or grease. This preparation will help the adhesive secure the panels to the surface.

Plan the installation of the panels before you begin. Use a laser level and mark vertical and horizontal lines on the walls as installation guides.

1.1.3 Trimming a Panel

When panels must be cut to fit an area, pull the fabric away from the panels, and then cut the panels to size with a sharp blade or carpet knife. Trim the fabric as necessary and reattach it to the panels by spraying adhesive on the back of the panels. You may need to trim a small wedge of fabric from the corner to get the fabric smoothly reattached to the panel.

When you must cut a hole in a panel to accommodate a fixture, be very careful not to cut the fabric at first when you are cutting the panel. To protect the fabric, pull it from one side of the panel, and then carefully place a piece of plastic, cardboard, or wood under the area to be cut. After cutting the desired shape from the panel, cut the fabric in an X-pattern and then smoothly fold the triangular-shaped pieces over the back of the panel. For circular cutouts, cut the fabric in two X-shaped patterns to create eight equal segments, as shown in *Figure 2*. You may want to apply a thin line of adhesive to the edge of fabrics that tend to unravel. When you are finished, attach the fabric to the back of the panel with spray adhesive.

NOTE

Sometimes a strip of plastic laminate is fastened with spray adhesive on the cut edge of the panel to sharpen the edge and make the fabric fold look more attractive.

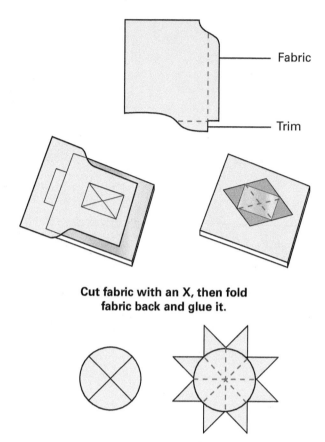

**Cut fabric with an X, then fold
fabric back and glue it.**

Figure 2 Trimming a panel.

1.1.4 Installation

Fiberglass ceiling panels are typically secured in the same manner as standard suspended ceiling tiles. (See NCCER Module 27209, *Suspended Ceilings and Acoustical Tile,* for more information.) Fiberglass wall panels can be installed in one or more of the following ways. Always be sure to follow the manufacturer's recommendation. Detailed installation drawings are often available to show installation of the panels.

Adhesives and Impaling Clips

Adhesives such as Liquid Nails® permanently secure ceiling and wall panels. Most adhesives set within 10 to 15 minutes, so you need to work quickly. Apply a bead of adhesive around the perimeter of the panels about 2 inches from the edges. Next, apply adhesive to the center of the panel in the shape of the letter S or W. Carefully set the panel in the desired location, and gently but firmly press it into place, leaving no gaps between the panel and the wall. Be sure to align the panel with the laser level lines. Small panels can be held in place for a few moments until the bond is secure. Larger panels will probably require some type of support. Impaling clips may be used for both fabric- and vinyl-covered panels. Nails will work for some fabric-covered ones.

During installation, impaling clips (*Figure 3*) help to support fiberglass panels until the adhesive sets. Impaling clips are a rectangular arrangement of sharp pins. The clips are attached to the wall with screws. The pins pierce the back of the fiberglass panel when it is positioned against the wall, holding it in place. The clips are not designed to permanently support the panels. When installing impaling clips, make sure that the screw is installed flush with the wall.

> **NOTE**
>
> Moisture in uncured concrete prevents adhesives from bonding to the surface. To use adhesive on uncured concrete surfaces, install furring strips to the wall and secure the adhesive to the strips.

3/4"

4"

Figure 3 Impaling clip.

> **WARNING!**
>
> Adhesives often contain ingredients that can cause eye, respiratory, and skin irritations. Always check the product's safety data sheet (SDS) for appropriate safety precautions. Be sure to use the proper personal protective equipment. Some adhesives outgas volatile organic compounds (VOCs), so be sure to use them with adequate ventilation.

Z-Clips

Z-Clips are attached to the back of panels, and then the panels are hung on a wall bar that is attached to the wall (*Figure 4*). There are some variations in the design of the bars. Some are attached together to the wall in an unbroken line. Another type of bar is called the wall clip, which is much shorter than the wall bar. The wall clip supports only one Z-Clip, so at least two clips must be installed per panel. Another style of bar is the double wall clip. The double wall clip is longer than a regular wall clip, and it supports two Z-Clips—one on each of two adjacent panels. Bars must be hung so they are level. Use a laser level and mark the wall at the correct level. Sometimes hook-and-loop fasteners are used with Z-Clips for added stability. Always consult the manufacturer's recommendations before using these fasteners.

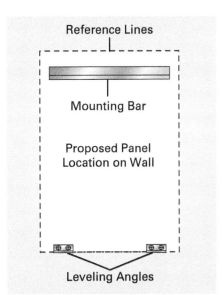
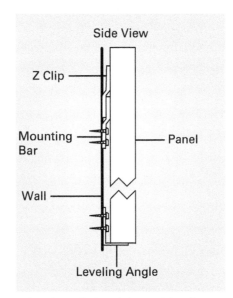

Note: The leveling angles help to ensure the panels are mounted evenly on a horizontal plane, and they also provide support for the weight of the panel.

Figure 4 Z Clips.

VELCRO® Hook-and-Loop Fasteners

These fasteners are often simply referred to as Velcro. They can be used to secure panels that need to be removable. They can also be used with other types of fasteners to provide additional support.

NOTE

VELCRO® brand hook-and-loop fasteners are available in regular and industrial strengths; never substitute one for the other.

Nails and Screws

Nails and screws can be used to secure panels to walls and ceilings. They are not usually a top choice for installation because they need to be sunk into the panels to hide them, which makes them very hard to remove. Also, they must be placed behind the fabric or disguised with a decorative cover. Finishing nails and screws can be placed and countersunk in panels covered with loosely woven fabric, but they can damage vinyl or finely woven fabrics.

Cleats

Cleats are thin strips of wood cut in symmetrical, trapezoidal shapes (*Figure 5*). One strip is mounted on the wall and another on the back of the panel. The panel is then hung on the cleat on the wall. Cleats allow panels to be easily removed without damage.

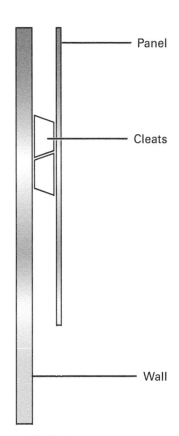

Figure 5 Cleats.

1.2.0   Wood Systems

Wood wall and ceiling systems are engineered for easy installation and maintenance. Most of these systems use wood **veneer** instead of solid wood for the finished surface. Veneer is a thin layer of wood that is glued to an engineered material. Veneers minimize the amount of wood used in the product, decreasing cost, weight, and expansion/contraction due to moisture content changes.

As with most commercial wall and ceiling systems, wood systems (*Figure 6*) are typically designed to improve the acoustical properties of an area. However, a smooth, flat surface can reflect sound waves in such a way that an echo is produced. To prevent this, the surface of these wood products is usually slotted, drilled with small holes, or grooved to change the way sound waves are reflected off it. In addition, these systems are typically backed with acoustical batting.

Figure 6 Wood wall system.
Source: Vladimir Nenezic/Shutterstock

1.2.1   Care

Wood can swell and shrink with increases and decreases in humidity. Because of this tendency, it is very important to expose the product to the climate of the area where it will be installed. The containers holding the product should be stored on the jobsite for a specified period of time. This storage period allows the product to adjust to the temperature and humidity of the area. If the containers are wrapped in plastic film, it may need to be removed prior to storage. Check the manufacturer's instructions for detailed information.

1.2.2   Installation

Wood panels are usually attached to walls with Z Clips, screws, or a combination of the two. Some wood panels are joined together by interlocking channels, such as tongue and groove joints (*Figure 7*).

Figure 7 Tongue and groove joint.

1.3.0   Glass Fiber Reinforced Gypsum

Glass fiber reinforced gypsum (GFRG or GRG) is a drywall product that has glass fibers mixed in it. Glass fibers give GFRG strength, so it can be thinner than standard drywall. It is also light in weight and easily molded into various shapes. These traits make GFRG desirable for ornate building features, such as columns, moldings, and arches. In the past, these features were created by skilled craftsmen using costly materials.

GFRG's light weight makes it useful for unique ceiling treatments, such as domed ceilings, light pockets, and vaulted ceilings (*Figure 8*). Properly primed GFRG can be finished with water-based paints, gold leaf, or stains. It can be finished to look like wood, metal, or marble. In addition, it is fireproof and will not be a source of combustion during a fire. The products must be primed as directed by the manufacturer and then finished as stated in the project specifications.

GFRG joints are finished using standard drywall procedures and supplies, such as tape and joint compound. Remember that GFRG is drywall and is subject to expansion and contraction with changes in humidity. When expansion joints are required, install them exactly as specified in the manufacturer's instructions.

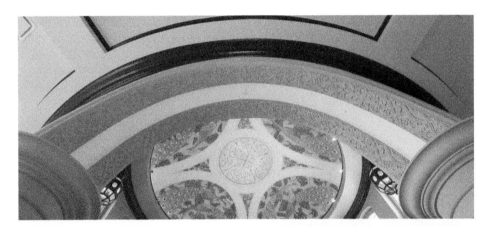

Figure 8 Examples of glass fiber reinforced gypsum products.
Source: Photos provided courtesy of Armstrong World Industries

1.3.1   Care

Because GFRG is drywall, it may only be used in areas with low humidity. By contrast, glass fiber reinforced concrete (GFRC) can be used in humid areas, such as fountains, pools, and other similar locations. For additional information about GFRC, consult NCCER Module 45205, *Exterior Cladding*.

When the shipping containers of GFRG arrive at the jobsite, carefully inspect them. Note any damage on the shipping invoice; if necessary, contact the manufacturer for instructions. Don't unpack containers unless instructed to do so by the manufacturer. An exception to this rule occurs when the purchase agreement states that products must be rejected on delivery. In this situation, unpack the product, carefully inspect it, and then repack it for storage. Store the products in a safe, dry location until they are installed. Never install damaged products.

1.3.2 Installation

Many GFRG products are custom-made for a specific location, so they may have unique installation instructions. Off-the-shelf items are typically designed to fit into standard framing. The following are general guidelines about GFRG products:

- Install products according to the manufacturer's instructions, using the recommended hardware and adhesives. Seal all joints with the recommended sealants.

- Keep a variety of shims on hand to help align the pieces as they are installed.

- When installing ceilings, use wire hangers and straps, support wires, and anchoring devices, just as with standard ceiling installations. Seismically active areas will require additional supports.

- Consult shop drawings for specific panel arrangement and layout.

1.4.0 Tectum® Panels

Tectum® products are made with wood fibers that are sprayed with a proprietary formula of concrete material and then molded into panels (*Figure 9*). These panels have a rough, irregular, fibrous surface, which gives it good acoustical properties. The material is durable and can withstand some abuse. It is often used on walls and ceilings in schools and gymnasiums. Because the wood fibers are coated with concrete, Tectum® panels are resistant to humidity and are often used near swimming pools. The panels cannot resist standing water, however.

Figure 9 Tectum® panels.
Source: Photos provided courtesy of Armstrong World Industries

Like standard acoustical ceiling panels, Tectum® ceiling panels are installed using lay-in, concealed tee, or profiled grid systems. Wall panels can be installed in the same manner as fabric- and vinyl-covered acoustical panels, except when used in high humidity areas. In these areas, the panels are usually installed on furring strips instead of flush with the wall. This installation allows the air to circulate behind the panel.

Tectum® products may be used with their natural finish, may be prefinished, or may be field painted, but painting them requires some caution. The wood fibers in the panels form random spaces and holes in the panels and on the surface. The spaces and holes on the surface give the material desirable acoustical properties. Paint can fill the spaces and change the acoustical properties of the material. Most Tectum® products can be painted with up to six layers of paint before their acoustical properties decline. When spray painting Tectum® products, be sure to use only enough paint to cover the fibers.

Tectum® roofing panels are often used for the upper deck of sports facilities (*Figure 10*). The panel shown in *Figure 11* combines an interior finish surface (Tectum® panel), insulation (expanded polystyrene), and OSB sheathing.

Going Green

Tectum® Products

An added benefit of Tectum® products is that they are biodegradable. These durable panels also contain no toxic binders, asbestos, or formaldehyde.

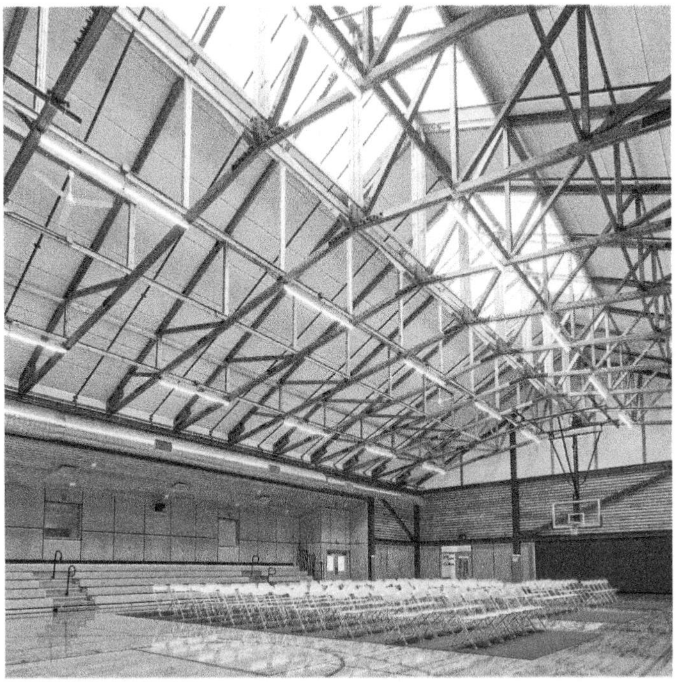

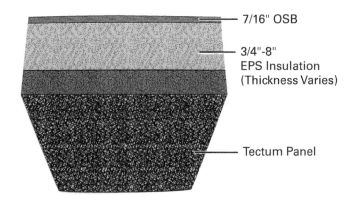

Figure 10 Tectum® roof deck.
Source: Photos provided courtesy of Armstrong World Industries

Figure 11 Tectum E composite panel.

1.5.0 Fabric Walls and Ceilings

Fabric walls and ceilings can add a distinctive finish to a commercial area. The fabric can be stretched over some type of framing system and then attached so that fasteners are invisible. This type of installation is also called wall upholstery. Fabric can be glued to the wall just like wallpaper. Fabric wall and ceiling coverings are light in weight and easy to install. When mounted on a framing system, the fabric can be easily removed and replaced. These coverings are often installed with an acoustical material backing to absorb sound.

When the fabric is glued to the wall, surface preparation includes patching any surface defects, and cleaning dust, dirt, grease, loose paint, and old wall coverings from the wall. If necessary, prime or paint the wall in a solid, neutral color and allow it to dry thoroughly before the fabric is applied. This step prevents color variations from being seen through the fabric. When recommended by the manufacturer, use a laser level and mark lines on the wall that can be used as guides when installing the fabric.

Some wall systems use proprietary framing systems. Be sure to follow the manufacturer's instructions. The following is a general procedure for installing a fabric wall system:

Step 1 Determine the desired position of the frame, and then use a level to mark vertical and horizontal lines that will be used as guides during installation.

Step 2 Measure and cut the frames to the desired lengths.

Step 3 Using the method recommended by the manufacturer, secure the frames to the wall on the lines snapped in Step 1.

Step 4 Apply acoustical material to the wall.

Step 5 Attach the fabric to the frame using the tool specified by the manufacturer. Be sure to keep the fabric straight and smooth.

1.6.0   Other Finishes and Treatments

This module has described interior treatments that greatly improve acoustical properties while beautifying a space. Some finishes are used purely to improve aesthetics.

1.6.1 Metal Column Covers

Commercial structures are built to be solid and durable. Sometimes the placement of structural components makes them appear unsightly. Many finish systems are designed to disguise these components. The metal columns shown in *Figure 12* are used to cover structural members such as I-beams. These covers can easily be attached using one of several different methods.

Figure 12 Metal column covers.
Source: idrisesen/Shutterstock

Metal column covers are available in modern and classical styles. The covers are designed strictly for looks and should never be used to carry a load. When installing the covers, handle them with care to avoid damaging the finish.

1.6.2 Reveal Trim

Reveal trim is used to finish the joints of drywall and acoustical tiles by creating a gap between the panels (*Figure 13*). This gap provides a quick and easy way to finish panel edges and create a visual feature. Reveal trim can be used on both ceilings and walls. It is often used to create a smooth transition between the wall and the ceiling. On walls, reveal trim may run horizontally or vertically.

1.6.3 CNC Router Work

A unique area of specialty interior treatments is the use of a computer numerical control (CNC) router (*Figure 14*). This machine is designed to cut precise, detailed shapes out of various materials including wood, plastic, high-density foam, some metals, and even sheetrock. Through the router's computer and software components, users design the specific cuts they want the machine to make. Cuts can be made in all directions, including right to left, front to back, up and down, and diagonal.

The applications for the CNC router are extremely varied. It can be used in carpentry, drywall installation, manufacturing, and many other areas. For specialty interior finishes, applications include creating ornate carvings in wood panels and doors; making clean, angled cuts in drywall that allow for complex assembly; and designing specialty moldings.

Figure 13 Reveal trim.
Source: Courtesy of Trim-Tex

Figure 14 CNC router.
Source: il21/Shutterstock

1.0.0 Section Review

1. True or False? When trimming a fabric panel, you must pull the fabric off the panel before cutting it to the appropriate size.
   a. True
   b. False

2. Wood can swell and shrink with increases and decreases in _____.
   a. daylight
   b. air pressure
   c. humidity
   d. heat

3. The glass fibers in GFRG give it strength, which means that compared to standard drywall, it can be made _____.
   a. heavier
   b. thinner
   c. thicker
   d. textured

4. When Tectum® is used in high humidity areas, the panels are usually installed on furring strips instead of flush with the wall to allow for _____.
   a. heat transfer
   b. a water buffer
   c. air circulation
   d. aesthetic spacing

5. True or False? Because fabric wall and ceiling coverings are mounted on a framing system, the fabric can be easily removed and replaced.
   a. True
   b. False

6. A CNC router can be used as a tool to make complex drywall assemblies because it can make _____.
   a. ornate carvings
   b. flashing
   c. drywall pieces
   d. angled cuts

2.0.0 Special Ceiling Systems

Performance Task

2. Describe in writing the parts and elements of a linear ceiling system.

Objective

Describe the specialty ceiling systems covered in this module, including their care and installation methods.

a. Explain the visual benefits of using a trim system on a suspended ceiling.
b. Describe the suspension system of a linear ceiling.
c. Understand that certain software programs allow contractors to easily design custom dome and vaulted ceilings.

A well-designed ceiling improves the aesthetics of a space by drawing the eye upward and creating visual interest. Ceiling islands, for example, create dimension and allow for special lighting designs. Wood slat systems add an impressive architectural element to an area, and domes and vaulted ceilings create a sense of grandeur.

The ceiling is also an important component of the acoustical properties of a space. It is often the largest flat surface in a room, which means that without some sort of sound-absorbent treatment, sound waves will reflect off of it, greatly increasing noise levels.

The ceiling systems discussed in this section address both the acoustics and the aesthetics of a space, with particular emphasis on visual impact. These systems require special installation procedures, but their care and handling are similar to the other wall and ceiling treatments described in this module. Protect the finishes from dirt and scratches; store all material in a clean, safe location; and never install any damaged product. Read on for more information on these impressive systems.

2.1.0 Trim Systems for Suspended Ceilings

Ceiling trim systems are used on suspended ceilings that are not attached to the walls. The trim is attached to the edge of the ceiling grids and panels to give them a finished look. In trim systems, suspended ceiling panels can be used to make unique ceiling features that diffuse overhead lighting, control acoustics, and add visual appeal to an area.

Trim systems come in many shapes and colors, and they can be used with a variety of ceiling systems. Always check the manufacturer's recommendations for ceiling panel and trim compatibility before you begin a job.

Trim system components are designed to be versatile and to minimize the number of cuts that need to be made in the field. Trim pieces may be straight or curved and may be mixed to create the desired shape. The Gordon VANTAGE POINT™ Contura perimeter trim shown in *Figure 15* uses straight trim pieces and 90-degree corners. For the most part, this type of design is a matter of correct assembly using splice plates, T-Bar connector clips, and corner hardware supplied by the manufacturer. The screws on these connectors need to be tightened until they are snug, but they should not be overtightened, which can warp the trim.

Compasso (TM) suspension trim can be shaped into many attractive designs. *Figure 16* shows examples of straight pieces of this trim. The curved pieces are a bit harder to install than the straight ones.

2.1.1 Installation

It's important to keep the following points in mind when you are installing ceiling trim systems:

- Study the reflected ceiling plan (RCP) or the manufacturer's shop drawings for information about the suspended ceiling design and arrangement.

- All suspended ceiling hardware must be attached to the upper deck of the area. This is especially important when an island is installed below a

Figure 15 Perimeter ceiling trim.
Source: Photograph Courtesy Gordon, Inc; www.gordon-inc.com

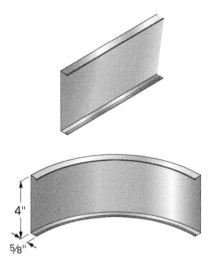

4"

5/8"

Figure 16 Compässo™ suspension trim (straight and curved).

suspended ceiling. Never secure an island's framework to a suspended ceiling unless it is recommended by the manufacturer.

- Ceiling trim is applied to the edges of suspended ceilings. The ceilings themselves must be installed in accordance with the appropriate local code, including seismic factors. When installing ceiling trim systems, use wire hangers and straps, support wires, and anchoring devices, just as with standard ceiling installations. Note that some trim systems require the trim to be secured to the upper deck with wire hangers, straps, or other anchoring devices. Be sure to check the manufacturer's instructions for details.

- Never cut any ceiling hanger devices in order to install a trim system. Doing so will weaken the structural integrity of the ceiling system.

- When ceiling islands are installed below a suspended ceiling, some equipment, such as fire sprinklers, must be installed through the island. Other equipment, such as lighting fixtures, may be installed either through the island or above it. Always check the RCP, project specifications, and construction plan detail drawings (if available). When fixtures are installed above islands, check to see if the upper side of the island's tiles will require some type of facing material. For example, when lighting is installed above an island, reflective material is often placed on the upper side of the tiles.

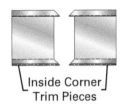

Inside Corner
Trim Pieces

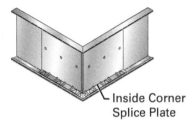

Inside Corner
Splice Plate

Figure 17 Compässo™ suspension trim (splice pieces).

- Always handle the trim carefully to avoid scratching the finish.
- Never mix pieces from different trim systems.
- Wash your hands before handling the trim pieces to avoid soiling the finish.

The procedure for placing a trim system will be somewhat different for each installation. The following steps may be used as a general installation guide:

Step 1 Install the ceiling grid to which the trim system will be attached. Make sure that the grid is larger than the island.

Step 2 Plan the trim installation by studying the RCP.

Step 3 Place the trim on top of the suspended grid.

Step 4 Using binder clips, fasten the trim pieces to the suspended grid in the desired shape. If necessary, place a paper towel or clean cloth over the trim to protect the finish before applying the clips.

Step 5 After all trim pieces have been laid out, use the appropriate splice connectors to join the trim pieces and form the desired shape (*Figure 17*). You will need to remove the binder clips to do this.

Step 6 Position the trim as desired on the grid and square the suspension grid as needed. Then reattach the binder clips.

Step 7 Mark the tees wherever they intersect with the trim.

Step 8 Using aviation snips, trim the tees as necessary.

Step 9 Fasten an attachment clip to the trim, and then insert the tee into the clip. Use a screw to attach the clip to the tee. Compässo™ trim pieces in widths of 10 inches and 12 inches require diagonal bracing with attachment clips.

Step 10 Cut the ceiling panels to fit, and then install them into the grid.

2.2.0 Linear Ceilings

Linear ceilings are lightweight strips of wood, or slats. The slats are clipped to a frame that is suspended from the ceiling by wires, rods, or some other type of hanging device. These systems may be fitted with lighting fixtures and heating/air conditioning vents. Some ceilings are backed with a sound absorbing material for noise reduction (*Figure 18*).

Figure 18 Wood linear ceilings.
Source: andy0man/Shutterstock

One disadvantage of wooden slats is that they can warp and twist. Warped or twisted slats cannot be installed because they will not seat properly. Before installation, each slat must be checked to make sure it is straight. This inspection can slow down the installation of a ceiling. One option is to separate unsuitable slats before starting the installation work. If this is done, the slats must be handled carefully to avoid damaging the finish. Return twisted or warped slats to the manufacturer for replacement.

2.2.1 Installation

As with all wall and ceiling systems, be sure to follow the manufacturer's instructions during installation of a linear ceiling. The following procedure may be used as a general installation guide:

Step 1   Study the reflected ceiling plan (RCP), the manufacturer's drawings, and/or the room finish schedules to find out the proper direction for orienting the wooden slats during installation, and to determine the ceiling elevation. When the area where the ceiling will be installed is multilevel, be sure to determine the ceiling elevation for each level. This is of special importance when the area has ramps and stairs.

Step 2   Using a laser level, mark a line at the desired ceiling elevation.

Step 3   Determine the level at which the ceiling perimeter trim should be installed. Depending on the system being used, you may need to add the thickness of the ceiling slat to the desired ceiling elevation to find the perimeter trim elevation. For example, if the ceiling slat is $\frac{3}{4}$" thick and the ceiling elevation is 11'-9", then the ceiling trim should be installed at 11'-9$\frac{3}{4}$". Check the manufacturer's instructions for more details.

Step 4   Install the ceiling perimeter trim according to the job specifications.

Step 5   Mark level lines at the desired location of the cliprails. In the example shown in *Figure 19*, the cliprails are installed 24" OC with a 4" allowance at the edges.

Step 6   Attach hanger clips to the ceiling in accordance with local code, and then fasten the hanger wires to the clips, as shown in *Figure 20*.

Step 7   Secure the cliprails to the hanger wire, ensuring that all cliprails are level. In this case, 12-gauge hanger wire is used. Mate the cliprail pieces together as necessary to reach the desired length.

Step 8   Examine each slat before you install it, making sure it is straight. Return twisted or warped slats to the manufacturer for replacement. Never install defective slats.

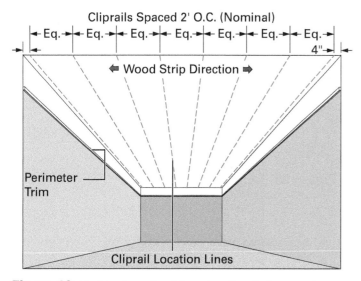

Figure 19 Marking laser level lines for cliprail placement.

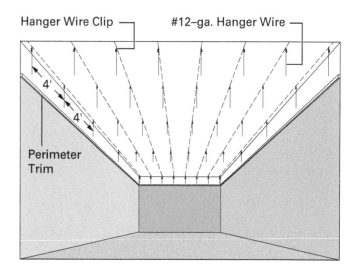

Figure 20 Suspending hanger wires.

Step 9 Starting at one end, snap the wooden slats into the connectors on the cliprail. The manufacturer may require the use of an installation tool. Some designs overlap, such as the one shown in *Figure 21*.

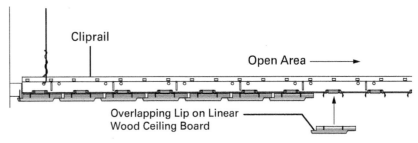

Figure 21 Installing slats onto the cliprail.

Step 10 If recommended by the manufacturer, secure the slat to the rail with screws placed at the recommended intervals (*Figure 22*).

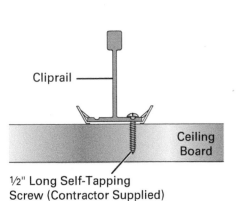

Figure 22 Securing the slat with screws.

2.3.0 | Domed and Vaulted Ceilings

Some companies, such as USG Corporation, sponsor computer software programs that allow contractors to easily design custom domes (*Figure 23*) and barrel vaulted ceilings (*Figure 24*). USG Corporation's software tool lets contractors fill in key information about the desired installation, and then it automatically designs the feature. The tool also creates customized installation instructions.

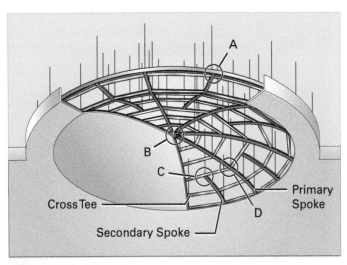

Figure 23 Dome feature.
Source: Courtesy of USG

Figure 24 Barrel vaulted ceiling.
Source: Courtesy of USG

The domes and vaults are created from components purchased from the manufacturer (*Figure 25*). The components are used to build a framework grid (*Figure 26*). The feature is hung into place and sheathed with drywall board and/or plaster compound. The surface can be finished in the same manner as any drywall product.

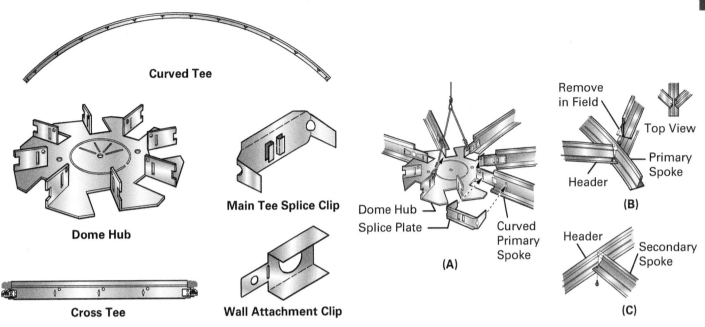

Figure 25 Dome components.

Figure 26 Dome framework grid.

2.0.0 Section Review

1. Ceiling trim systems are used on suspended ceilings that are not attached to the _____.
   a. ceiling panels
   b. walls
   c. drywall
   d. ceiling joists

2. Linear ceilings are lightweight strips of wood, or slats, that are clipped to a frame that is suspended from the ceiling by _____.
   a. ropes
   b. chains
   c. screws
   d. wires

3. After a contractor designs a custom ceiling dome, the components are made by a manufacturer and installed in a _____.
   a. specialized pattern
   b. framework grid
   c. circular method
   d. numbered series

	# 3.0.0 Additional Safety Hazards
Performance Task	## Objective
3. Review the safety data sheet (SDS) of a common adhesive or solvent used for interior finish systems. Identify the hazards of the material and mitigation measures (e.g., using a ventilator).	Describe additional safety hazards associated with interior specialties. a. Identify key hazards associated with installing different types of interior specialties.

Interior systems installed above ground level require specific safety procedures. Special care must be taken when ladders and scaffolding or when using interior finishing products.

3.1.0 Safety When Working at Heights

When installing interior specialties, there are three primary areas that pose serious hazards. These areas include:

- Working at heights
- Scaffolding
- Products and chemicals

3.1.1 Aboveground Safety

Working even a few feet above ground adds an element of danger to the job. Use the correct height of equipment for each job and resist the urge to over-reach for your equipment. Be sure that all equipment is in good working order. Any defective items must be tagged immediately so that they will not be used. Exercise caution when handling sharp tools and metal components.

When working above ground, wear the appropriate personal protective equipment, maintain control of materials and tools at all times, and stay alert for personnel working on the ground below. If you are working on the ground when work is being done overhead, wear the appropriate personal protective equipment and stay alert for falling debris and tools.

3.1.2 Scaffolding

Doing overhead work often requires the use of scaffolding or ladders to reach the work area. Scaffolding must be inspected by a competent person before it is used on each shift and whenever it is moved or modified in any way. Some scaffolding must be designed by an engineer, so know the practices at your worksite. If you are unsure of the site requirements, ask your supervisor before you begin work.

3.1.3 Products and Chemicals

When using any product, first check its safety data sheet (SDS). Use the recommended safety precautions. Never smoke where adhesives, paints, solvents, or other chemicals are being used. Wash your hands before eating, before and after using the rest room, and when you are leaving the worksite.

3.0.0 Section Review

1. True or False? If you're working just a few feet off of the ground, it's safe to over-reach for your equipment.
   a. True
   b. False

2. Scaffolding must be inspected by a competent person before it is used on each shift and whenever it is moved or _____.
   a. made out of metal
   b. modified in any way
   c. used by two or more people
   d. splashed with materials

Module 45204 Review Questions

1. The best way to store fabric-covered fiberglass wall panels is _____.
   a. upright in a storage room inside the building
   b. flat and covered in a storage room inside the building
   c. flat and covered anyplace outside the building
   d. upright anyplace outside the building

2. True or False? After trimming the fabric on a fabric-covered panel, Liquid Nails® adhesive should be used to reattach the fabric to the panel.
   a. True
   b. False

3. When a square hole for a light fixture must be cut in a fabric-covered fiberglass panel, the type of cut to make in the fabric before folding it back is _____.
   a. V-shaped
   b. X-shaped
   c. W-shaped
   d. Y-shaped

4. Adhesives that are used to secure fiberglass panels to walls set in about _____.
   a. 10 to 15 minutes
   b. 30 to 60 minutes
   c. 1 to 4 hours
   d. 24 hours

5. The screws that attach impaling clips to the wall should _____.
   a. be coated with adhesive
   b. protrude slightly from the wall
   c. be disguised with a decorative cover
   d. be installed flush with the wall

6. Screws can be used to mount fiberglass panels that are covered with _____.
   a. loosely woven vinyl
   b. loosely woven fabric
   c. tightly woven vinyl
   d. tightly woven fabric

7. To decrease costs and weight, most wood ceilings and wall systems use _____.
   a. slats
   b. planks
   c. veneers
   d. panels

8. The acoustical properties of wood wall panels are improved by _____.
   a. installing them with spacers between them
   b. drilling holes or cutting slots in them
   c. sanding them
   d. painting them

9. True or False? The glass fibers in glass fiber reinforced gypsum (GFRG) prevent the expansion and contraction that can occur with changes in humidity.
   a. True
   b. False

10. True or False? If properly primed, GFRG can be finished with various materials, including water-based paints, gold leaf, and stains.
    a. True
    b. False

11. Tectum® products are made with _____.
    a. wood fibers sprayed with concrete material
    b. glass fibers mixed with adhesives
    c. fabric fibers
    d. fiberglass

12. True or False? Tectum® panels can be installed near swimming pools.
    a. True
    b. False

13. Before installing the framework for upholstered walls, you should _____.
    a. scrape old paint from the wall
    b. tape the fabric to the wall to make sure that it fits properly
    c. thoroughly wash the walls
    d. mark horizontal and vertical laser level lines to be used as guides

14. When installing trim systems on a suspended ceiling island, _____.
    a. secure the island hangers to the suspended ceiling above it
    b. install the fire sprinklers through the island
    c. be careful not to splash the panels when you paint the trim
    d. it is okay to cut a hanger device, if necessary

15. To protect the finish on the pieces of a trim system, you should _____.
    a. dust your hands with talcum powder before handling trim
    b. wash your hands before handling trim
    c. keep the trim covered during installation
    d. firmly place binder clips so the trim will not slip during installation

16. When installing trim systems on a suspended ceiling island, trim the tees with _____.
    a. a circular saw
    b. power scissors
    c. aviation snips
    d. GFRG

17. When installing linear ceilings, secure the cliprails to the _____.
    a. hanger wire
    b. hanger wire clip
    c. laser level lines
    d. sheetrock

18. When using any product or chemical, first check its _____.
    a. ingredient label
    b. where it was manufactured
    c. company website
    d. safety data sheet

Answers to odd-numbered questions are found in the Module Review Answer Key at the back of this book.

Answers to Section Review Questions

Answer	Section	Objective
Section One		
1. a	1.1.3	1a
2. c	1.2.1	1b
3. b	1.3.0	1c
4. c	1.4.0	1d
5. a	1.5.0	1e
6. d	1.6.3	1f
Section Two		
1. b	2.1.0	2a
2. d	2.2.0	2b
3. b	2.3.0	2c
Section Three		
1. b	3.1.1	3a
2. b	3.1.2	3a

Exterior Cladding

Objectives

Successful completion of this module prepares you to do the following:

1. Explain the purpose of building wraps and why they're usually the first step in exterior cladding.
   a. Explain why building wraps have largely replaced building paper.
   b. Describe the basic makeup of building wraps and the sizes they come in.
   c. Describe the methods of installing and securing wrap.
2. Identify the basic components of trim and their respective functions.
   a. Explain the purpose of control joints, weep screed, starter joints, and flashing.
3. Identify and describe the different types of exterior cladding systems.
   a. Outline the process of installing stucco.
   b. Describe the benefits of using synthetic stone instead of real stone.
   c. Outline the process of installing water managed EIFS.
   d. Understand the advantages of fiber-cement siding over wood siding and the different shapes and styles it comes in.
   e. Describe building features commonly created with glass fiber reinforced concrete (GFRC).

Performance Tasks

Under supervision, you should be able to do the following:

1. On site, observe a building wrap cut around an opening and evaluate whether it has been executed properly for drainage.
2. Observe samples of control joints, weep screed, starter strips, and flashing and compare and contrast them.
3. Observe a detailed figure of an EIFS and identify the designed path of moisture drainage (where it enters, where it drains, where it exits the system, etc.).

Overview

Exterior cladding places the finishing touches on a building, and it protects a building's structural members from the damaging effects of weather. This module explains the similarities and differences between commonly used exterior claddings. In addition, it covers the proper installation procedures that should be followed to ensure that cladding protects the structural integrity of a building.

Digital Resources for Drywall

Scan this code using the camera on your phone or mobile device to view the digital resources related to this craft.

1.0.0 Building Wraps

Performance Task

1. On site, observe a building wrap cut around an opening and evaluate whether it has been executed properly for drainage.

Objective

Explain the purpose of building wraps and why they're usually the first step in exterior cladding.

a. Explain why building wraps have largely replaced building paper.

b. Describe the basic makeup of building wraps and the sizes they come in.

c. Describe the methods of installing and securing wrap.

Exterior cladding improves a building's appearance, but that is not its primary purpose. In fact, one of its main functions is to protect buildings from the harsh and damaging effects of weather. Another is to help provide the necessary air, moisture, and temperature barrier between a building's interior and exterior.

While not always required by the local code, a properly applied building wrap gives a home or building a weather-resistant seal that will block hot and cold air and help lower heating and cooling costs. It also acts as a moisture barrier and helps to keep water from damaging the exterior frame. These properties often make the building wrap the first step of the exterior cladding process (*Figure 1*).

Residential Application   Commercial Application

Figure 1 Building wrap.
Source: (left) © 2024 James Hardie Building Products Inc. All Rights Reserved.™; (right) Scott Ehardt via Wikimedia

For building wrap to be effective, all gaps must be secured properly with tape approved by the manufacturer. Gaps include spaces around doors and windows, as well as any tears in the wrap. The covering also needs to be properly designed and installed in order to prevent the introduction of moisture into building cavities, which can result in the growth of mold.

1.1.0 Building Paper Versus Building Wrap

Traditionally, structures have been covered with water-resistant building paper. This paper helps prevent water that may have leaked through the primary barrier (siding) from reaching the structural sheathing or other components of the structure. The paper is water **permeable** so that moisture inside the walls can pass through and evaporate. To some extent, the paper reduces air infiltration of the structure, especially when board sheathing is used.

House wraps or building wraps are now largely used in place of building paper. These products, with brand names including Tyvek®, HardieWrap®, or PermaPro®, are easier to apply and perform the same functions as building paper. When properly applied and sealed, the wraps help create a nearly airtight structure no matter what sheathing material is used. Most versions of these wraps are an excellent secondary barrier under all siding, including stucco and exterior insulation and finish systems (EIFS).

Permeable: Having pores or openings that permit liquids, such as water, or gases to pass through.

1.2.0 Building Wrap Systems

Most building wrap is made of spun, high-density polyethylene fibers randomly bonded into an extremely tough, durable sheet material. Building wrap is usually available in several versions and weights for residential and light commercial use. Special versions are available with vertical water channels permanently pressed into the material for stucco and EIFS. Building wrap is available in rolls ranging in widths from 18 inches to 10 feet, and in lengths from 100 to 200 feet.

> **WARNING!**
>
> Some building wraps are slippery and should not be used in any application where people might walk on them. Because the surface is slippery, use pump jacks or scaffolding for exterior work above the lower floor. If ladders are used, extra precautions must be taken to prevent them from sliding on the wrap.

1.3.0 Securing the Wrap

There are various ways to secure the wrap. Common methods include nails with large heads, nails or screws with plastic washers (*Figure 2*), or 1-inch-wide staples, depending on the manufacturer's instructions. Screws and washers are used for steel construction. Special contractor's tape (*Figure 3*) or sealants compatible with the wrap are used to seal the edges and joints.

Figure 2 Screws with plastic washers for securing building wrap.
Source: Courtesy of GAF

Figure 3 Contractor's tape.

When installing building wrap, always refer to the manufacturer's instructions for specific installation information. In general terms, building wrap is installed as follows:

Step 1   Using two people and beginning at a corner on one side of the structure, leave 6" to 12" of the wrap extended beyond the corner to be used as an overlap on the adjacent side of the structure (*Figure 4*). Align the roll vertically and unroll it for a short distance. Check that the stud marks on the wrap align with the studs of the structure. Also check that the bottom edge of the wrap extends over the foundation line by 1" and runs along the foundation line. Secure the wrap to the corner at 12" to 18" intervals.

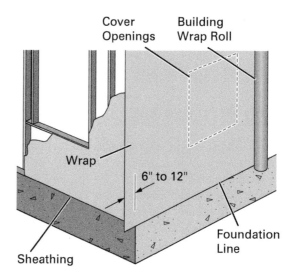

Figure 4 Starting a roll of building wrap.

Step 2   Unroll the wrap two or three more feet and ensure that it overlaps and runs along the foundation line. Secure the wrap vertically at 12" to 18" intervals on each stud using the stud marks as a fastening guide. Continue around the structure, covering all openings. If a new roll is started, overlap the end of the previous roll 6" to 12" to align the stud marks of the new roll with the studs of the structure.

Step 3   If the upper floors or parts of the structure require coverage, repeat Steps 1 and 2, starting above the existing wrap. Make sure that the bottom edge of the upper layer of wrap overlaps the top edge of the lower layer 6" to 12".

Step 4   At the top plate, ensure the wrap covers both members of the double top plate (*Figure 5*), but leave the flap loose for the time being.

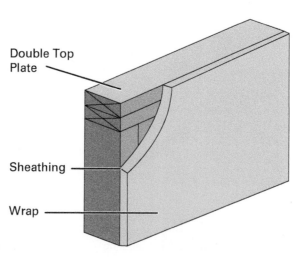

Figure 5 Top plate detail.

Step 5 Depending on whether the windows and doors are uninstalled or installed, use one of the following two methods to cut back the wrap:

Method 1—Uninstalled windows/doors:

- At the opening, cut the wrap as shown in *Figure 6*. Fold the three flaps around the sides and bottom of the opening and secure every 6 inches. Trim off the excess.

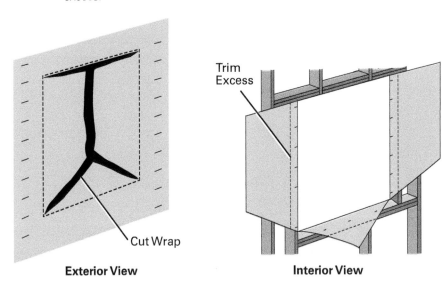

Exterior View   **Interior View**

Figure 6 Cutting and folding building wrap at a rough opening.

- Along the outside, install 6" **flashing** paper at the bottom of the opening, then up the sides over the top of the wrap.
- Install head flashing at the top of the opening under the wrap and over the side flashing (*Figure 7*). Tape the flap ends to the head flashing using manufacturer-approved tape. If a metal flashing is required, as shown in *Figure 7*, then install it on top of the flashing paper at the top of the window.
- Install the wrap at the head and **jambs** in shingle fashion, lapping as shown in *Figure 8*.

Flashing: Thin strips of material that are placed around openings in a structure to divert water away from the opening.

Jambs: The top and sides of a door or window frame that are in contact with the door or sash.

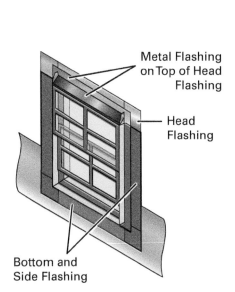

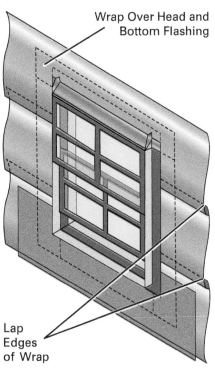

Figure 7 Installing flashing around an opening.

Figure 8 Lapping the wrap in shingle fashion.

All flashing must be designed and installed to shed moisture. Failure to do so will result in moisture buildup, possible mold growth, and building damage.

Method 2 — Installed windows/doors with flanges:

- With the house wrap installed, create a top flap of the wrap. Insert head flashing under the flap and over the flange.
- Extend the flashing to the sides about 4" and tape the flap to the head flashing.
- On the remaining sides, trim the wrap to overlap the flange area and tape the edge to the flanges (*Figure 9*).

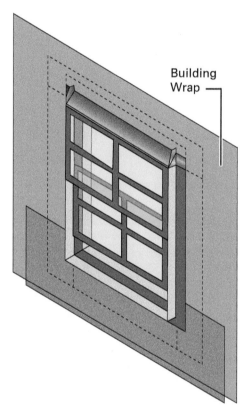

Building
Wrap

Figure 9 Installing wrap with a window in place.

Step 6   Secure all the bottom edges of the wrap to the foundation with the manufacturer-recommended joint sealant, then fasten the lower edge to the sill plate. At the top plate, seal the edge to the upper plate with the sealant and fasten the edge to the plate.

Step 7   Seal all vertical and horizontal joints in the wrap with the recommended tape.

Step 8   Before applying the siding, repair any damage or tears in the wrap with tape or sealant.

Always follow the window or door manufacturer's recommendations for the installation of flashing around windows or doors.

1.0.0 Section Review

1. The main difference between building paper and building wrap is that the wrap creates a nearly airtight structure whereas the paper is water _____.
   a. proof
   b. resistant
   c. permeable
   d. tight

2. Most building wrap is made of spun, high-density _____.
   a. polyethylene fibers
   b. glass fibers
   c. cement fibers
   d. paper fibers

3. True or False? A different method of installing building wrap should be used if windows and doors have already been installed.
   a. True
   b. False

2.0.0 Trim

Objective	Performance Task
Identify the basic components of trim and their respective functions. a. Explain the purpose of control joints, weep screed, starter joints, and flashing.	2. Observe samples of control joints, weep screed, starter strips, and flashing and compare and contrast them.

Trim is an important component of exterior cladding. Omitting or improperly installing even small pieces of trim or forgetting to apply the correct sealant to openings can cause the cladding to fail.

2.1.0 Types of Trim

Often a variety of trim pieces are needed to properly install exterior cladding. Manufacturers frequently make proprietary trim to accompany their cladding systems. When this is the case, it is very important to use the manufacturer's recommended trim. This section covers the basic trim pieces and their functions.

2.1.1 Control Joints

Control joints are used with stucco and exterior insulation and finish systems (EIFS) to relieve stress in the coating due to thermal expansion and contraction. Control joints also help to control cracks caused by these thermal changes (*Figure 10*). With conventional stucco, the control joints must be designed to coordinate with the drainage system, with particular attention paid to splices and intersections.

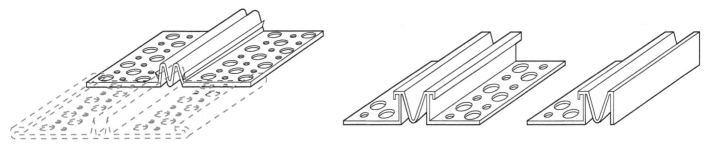

Figure 10 Control joints.

2.1.2 Weep Screed

Weep screed is used with stucco and other cladding systems to drain water from behind the cladding (*Figure 11*). It's most commonly installed at the base of the exterior wall and is designed to allow water to "weep" through the holes of its v-shaped extension, encouraging proper drainage.

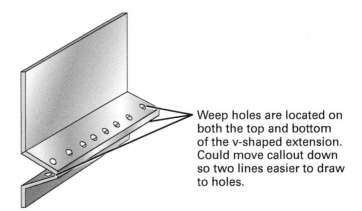

Weep holes are located on both the top and bottom of the v-shaped extension. Could move callout down so two lines easier to draw to holes.

Figure 11 Weep screed.

2.1.3 Starter Strips

Starter strips (*Figure 12*) are used with EIFS to support the insulated panel and to allow water to drain from behind the cladding. Other cladding systems also use starter strips, but the design is different. When installing a starter strip, use a level and mark a line at the installation level, and then install the starter strip on that line.

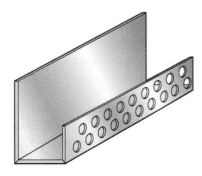

Figure 12 Starter strip.

2.1.4 Flashing

Flashing is used with all cladding systems (*Figure 13*). Flashings are thin pieces of metal or flexible rubberized material that are installed above openings in the cladding system to prevent water from seeping behind the cladding. Flashings are installed at a slope that causes water to flow away from the opening. Improperly installed or omitted flashing is one of the causes of cladding failure.

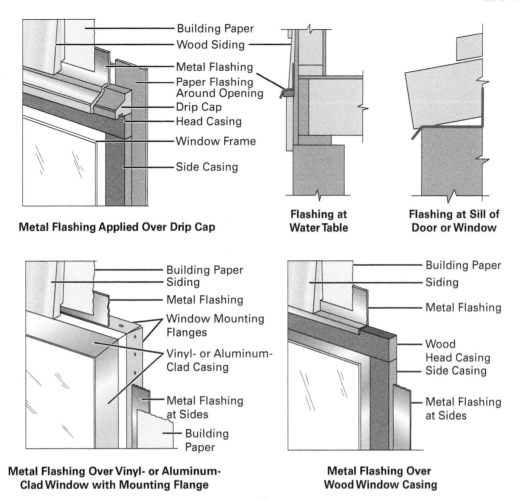

Metal Flashing Applied Over Drip Cap

- Building Paper
- Wood Siding
- Metal Flashing
- Paper Flashing Around Opening
- Drip Cap
- Head Casing
- Window Frame
- Side Casing

Flashing at Water Table

Flashing at Sill of Door or Window

Metal Flashing Over Vinyl- or Aluminum-Clad Window with Mounting Flange

- Building Paper
- Siding
- Metal Flashing
- Window Mounting Flanges
- Vinyl- or Aluminum-Clad Casing
- Metal Flashing at Sides
- Building Paper

Metal Flashing Over Wood Window Casing

- Building Paper
- Siding
- Metal Flashing
- Wood Head Casing
- Side Casing
- Metal Flashing at Sides

Figure 13 Flashing.

2.0.0 Section Review

1. Control joints are used with stucco and EIFS to relieve stress in the coating due to _____.
   a. water retention
   b. thermal expansion and contraction
   c. insect activity
   d. wet and dry cycles

2. Weep screed is most commonly installed at the base of the _____.
   a. foundation
   b. cladding
   c. substrate
   d. exterior wall

3. Starter strips are used with EIFS to support the _____.
   a. wrapped mesh
   b. insulated panel
   c. exterior wall
   d. system weight

4. One of the causes of cladding failure is improperly installed or omitted _____.
   a. flashing
   b. webbing
   c. screed
   d. siding

3.0.0 Exterior Cladding Systems

Performance Task

3. Observe a detailed figure of an EIFS and identify the designed path of moisture drainage (where it enters, where it drains, where it exits the system, etc.).

Objective

Identify and describe the different types of exterior cladding systems.

a. Outline the process of installing stucco.
b. Describe the benefits of using synthetic stone instead of real stone.
c. Outline the process of installing water managed EIFS.

d. Understand the advantages of fiber-cement siding over wood siding and the different shapes and styles it comes in.
e. Describe building features commonly created with glass fiber reinforced concrete (GFRC).

Exterior cladding does more than make a building look good. Its primary purpose is to protect the structural members of a building from air and water infiltration. For exterior cladding to do its job, it must be installed properly. Improperly installed cladding can make the building's occupants uncomfortable due to drafts and temperature changes. It can also seriously damage the building's structural soundness.

This section provides information about some of the most common types of exterior cladding used today, including stucco, synthetic stone, exterior insulation and finish systems (EIFS), fiber-cement siding, and glass fiber reinforced concrete (GFRC). You'll also learn proper installation techniques for all of these types of cladding.

3.1.0 Stucco

Traditional stucco is a durable, cement-based coating for exterior walls that is breathable and allows moisture to pass through. Acrylic stucco is a synthetic material that provides a waterproof finish. Stucco can be applied over wood sheathing, fiberglass mat gypsum, or cement board, as well as over cement block and poured concrete walls. When applying stucco, building wrap, such as Tyvek®, Commercial Wrap®, or HomeWrap® must be placed over the sheathing of wood frame structures. Building wrap is not needed over cement block or poured concrete walls.

The wrap is covered with 27" × 96" pieces of metal or vinyl reinforcement lath as recommended by the manufacturer of the stucco. The lath is applied using special spacing fasteners or on top of a plastic mesh material that holds it slightly away from the wall. This ensures a space for water drainage.

Traditional and acrylic stucco finishes tend to expand and contract with temperature variations. These changes can cause cracks in the surface, so expansion and control joints are used. It is important to understand that these joints do not prevent cracks; they merely control their location and shape. Construction documents will specify the placement of expansion and control joints. Typically, they are placed at corners, around openings such as doors and windows, and at lath joints. Placing expansion and contraction joints near doors and windows helps to disguise cracks. This is because the door or window breaks the expanse of stucco, and the joints can be hidden by trim.

For a stucco finish, three coats of cement plaster must be applied (*Figure 14*). The first coat, called the *scratch coat*, is applied about $\frac{1}{4}$" to $\frac{1}{2}$" thick over the lath. After the scratch coat sets, but has not hardened, it is scratched with a scratch rake. The scratches should be about $\frac{1}{8}$" deep. The scratches help the next layer adhere to the scratch coat. The scratch coat must be allowed to harden according to the manufacturer's specifications (usually 36 to 48 hours).

The next coat is called the *brown coat*. Before applying this coat, the scratch coat must be sprayed with clean water. This helps the brown coat adhere to the scratch coat. The first two coats can be applied with a rough finish using a trowel.

NOTE

All control joints must be properly flashed as a part of the complete drainage system.

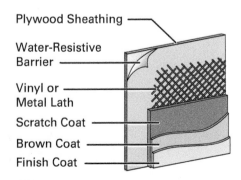

Plywood Sheathing

Water-Resistive Barrier

Vinyl or Metal Lath

Scratch Coat

Brown Coat

Finish Coat

Figure 14 Stucco section.

Stucco Drying Time

Stucco must be allowed to dry between coats. The amount of drying time required depends on the environmental conditions at the time the coats are applied. Refer to the manufacturer's instructions for the recommended drying/curing times.

The final coat, or *finish coat*, can be applied rough or smooth as desired using a variety of finishing techniques. The total thickness of the three coats should be about $\frac{1}{2}$" to $\frac{5}{8}$" with no lath and $\frac{3}{4}$" to 1" with lath, according to the Stucco Manufacturers Association.

The scratch, brown, and finish coats are a mixture of cement, sand, and clean water in different proportions. The Standard Specification for Application of Portland Cement-Based Plaster Proportions (ASTM C926) defines the following proportions for the coats:

- *Scratch coat* — 1 part cement to $2\frac{1}{4}$ to 4 parts sand
- *Brown coat* — 1 part cement to 3 to 5 parts sand
- *Finish coat* — 1 part cement to $1\frac{1}{2}$ to 3 parts sand

Wood trim, such as frieze board or half timbers, is sometimes applied next to stucco to simulate a particular style of architecture. In such applications, the wood trim should have a rabbet on the back joining edge, called a *stucco lock* (*Figure 15*). When properly designed and installed, the stucco lock prevents water penetration around the stucco at the point where it joins the wood trim. When applying stucco to concrete block walls, a water barrier and lath are not necessary, but all other requirements apply.

Stucco needs to be protected from freezing for the first 48 hours after application because the water in it can freeze and cause cracks in the finish. Stucco should be cured according to manufacturer's instructions, usually at temperatures above 40°F. In addition, recently applied stucco may dry out too quickly in conditions of high temperature, low humidity, and wind.

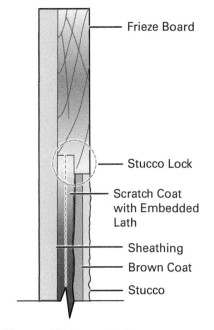

Figure 15 Stucco lock.

3.2.0   Synthetic Stone

Synthetic stone is often used in place of real stone (*Figure 16*). Synthetic stone looks like and is installed like real stone. However, working with synthetic stone avoids many of the difficulties of real stone. Real stones are heavy; are different in color, size, and shape; and can be smooth or jagged. Typically, real stones must be trimmed and smoothed before they can be used as cladding.

Figure 16 Synthetic stone.
Source: Compliments of Acme Brick Company

Synthetic stone comes in many different styles, such as fieldstone and ledge-stone. This variety allows designers to achieve almost any desired look: rustic, contemporary, traditional, or elegant. Synthetic stone is an attractive finish for frame or concrete structures, and it can be used on exteriors and interiors.

Synthetic stones are manufactured to exact specifications using lightweight concrete and high-density polyurethane. While the 'stones' appear random in size, shape, and color, they are actually engineered to look like real stones, but are easier to install than real stones.

Wall preparation for the installation of synthetic stone is typically the same as for stucco. Some types of synthetic stones are fastened to the wall with hardware or adhesive. Other types are placed with grout, similar to real stones. Once all stones have been set, the spaces between them are filled with grout. As with all cladding, it is very important to seal any gaps around openings, such as doors and windows, to prevent water leaks.

Installing stonework is not as fast as other cladding methods. It takes a great deal of careful preparation. Before you begin installation, it's very important to figure out an appropriate pattern and layout for the stones. The following are helpful procedures to get you started:

Step 1   Remove all of the stones from their boxes and lay them out on the ground in a clean area near the installation site. If you are working in an area where the stones can get dirty, set them on scrap plywood or another material.

Step 2   Mix the stones from different boxes so that the variations are randomly dispersed.

Step 3   Lay out the stones in an attractive pattern, mixing sizes, colors, and shapes. Use the following guidelines:

- Leave no more than a $1/2$" gap between stones.
- Avoid placing stones in ways that will require long grout lines. Use different-sized stones to break up patterns.
- At inside corners, alternate placing stones on each wall, so they can be woven together. Use an alternating long-short pattern, cutting the stones as needed to ensure a proper fit. Be sure to cut the stone so that the cut side is facing the corner.
- Determine if stones and grout are to be sealed. Sometimes sealing is required if the stone will be in an area where soil can be splashed onto its surface, such as near a flower bed. Always keep in mind that sealing the stones and grout can darken the color.
- Some manufacturers make trim pieces that can be used on outside corners and around doors and windows. When using these pieces, place them first, and then place the stones to the trim. Use small stones to avoid having to cut pieces to fit.

Sometimes you'll need to cut a stone to fit it properly. In that case, check the manufacturer's recommendations about how to cut it and what personal protective equipment you should use. Additionally, keep the following guidelines in mind:

- Be sure to clean all dust from the stone before you set it. Any dust on the stone can prevent a good bond with the grout.
- Always set a cut stone in a way that the cut side is not visible. That is, when the stone is above eye level, set the cut side facing up; when the stone is below eye level, set the cut side facing down.
- Do not cut stones that will be set at eye level.

When you are ready to place the stones, secure them according to the manufacturer's specification. The following procedure describes setting stones with grout:

Step 1   Mix the grout according to the manufacturer's instructions, adding coloring as desired. Always mix the grout and coloring agent in the same proportions to avoid different shades of color. Mix only the quantity of grout that you can use before it sets. Never use partially set grout. This can cause the installation to fail.

Step 2 When instructed to do so by the manufacturer, moisten the brown coat with clean water. This helps the grout to adhere to the base coat. This process may need to be repeated periodically when the weather is warm and dry or when the installation area is in direct sunlight.

Step 3 Determine the direction in which the stones will be installed. Some products need to be installed from the top down. This helps to keep the stones clean of spattered grout. Some products should be installed from the bottom up. This helps to keep the stone line level.

Step 4 Once you have established a starting point, use a level and mark a line at the starting level. When installing flat stones, such as brick and ledgestone, mark lines every 8" to 12" above the starting level. This will help to keep a level installation.

Step 5 Dust off the stones using a clean whisk broom. Always clean all dirt, sand, and dust from the stones before placing them. This will ensure a good bond with the grout.

Step 6 Using a metal trowel, spread about $\frac{1}{2}$" of grout on the back of the stone. Set the stone roughly $\frac{1}{2}$" from its neighboring stones. Firmly press the stone into place until grout oozes from the back of it. Firm placement is needed to get a good bond. An alternate method is to spread about $\frac{1}{2}$" of grout over the brown coat, and then firmly place the stones until the area is covered. When using this method, spread the grout over a small area of roughly 4' by 4'. This technique helps prevent the grout from drying out before the stones are set.

NOTE

Never try to wipe away wet grout from synthetic stone. This can smear the grout and permanently stain the stone. Instead, wait until the grout is dry, and then brush it away with a clean whisk broom.

Some manufacturers recommend dampening the back of the stone to help the grout adhere to it. When this is the case, be sure that the stone is not so wet that water puddles on its surface. Excess water will prevent the grout from adhering to the stone.

Step 7 Once all stone in the area has been set, use a caulk gun, or fill a grout bag about two-thirds full of grout and twist the top to keep it closed. Carefully squeeze the grout into the cracks between the stones. Take care that the joint line is free of gaps and the grout is free of air bubbles.

Step 8 Allow the grout to firm but not set, and then finish the grout with a striking tool to remove any excess, leaving a concave surface.

Step 9 Seal the grout and stones if this is recommended by the manufacturer.

3.3.0   Exterior Insulation and Finish Systems (EIFS)

Exterior insulation and finish systems (or EIFS) combine insulation with a synthetic, textured exterior coating. This finishing system brings together the following: a water-resistive barrier (WRB); a drainage plane; an insulating panel of expanded polystyrene; mesh; a water-resistive base coat that acts as a weather barrier; and a crack-resistant finish coat. The finish coat, called synthetic stucco, is an elastomeric coating. This cladding looks attractive and provides good insulation from both heat and cold, which helps lower heating and cooling costs (*Figure 17*).

There are two types of EIFS: traditional and water managed. Traditional systems relied on the insulation and coating to prevent water infiltration, but this expectation is unrealistic. Even condensation from thermal changes can cause water to collect behind the panel, causing wood surfaces to rot and encouraging mold to grow. Today, water managed systems are used instead. The latest system uses water barriers to prevent leaks.

EIFS is a multilayer system similar to those used in stucco finishes (*Figure 18*). The most critical layer is the water management layer. Wood-framed buildings are covered with an appropriate moisture barrier. Likewise, any walls with exterior wood, including masonry or concrete buildings with wood furring, must

Figure 17 EIFS.
Source: s74/Shutterstock

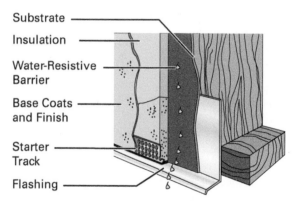

Figure 18 Typical EIFS installation.

be coated with a water-resistive barrier (WRB) to prevent water from contacting the wood. In addition, flashing must be installed over doors, windows, and any other opening that can allow water to enter. Instead of seeping into the wall, the water is drained out the bottom of the system. For this process to be effective, there can be no tears, openings, or breaks in the barrier.

Proper installation of EIFS is critical to ensure that the water managed system works effectively. It is extremely important to follow the manufacturer's instructions when installing EIFS to prevent voiding any warranties. Use only manufacturer-recommended barriers, adhesives, and components.

After the WRB is installed properly, a layer of insulation is applied. Panels of expanded polystyrene are usually used as the insulation in this system. Polystyrene is commonly used as building insulation. Expanded polystyrene (EPS) is made from small pellets of polystyrene that have been molded together into a particular shape—panels, in this case. Extruded polystyrene (XPS) is a foam of polystyrene that is pressed into a mold. XPS is sold under the brand name Styrofoam™. The term Styrofoam™ is often mistakenly used to describe EPS. When polystyrene is made up of pellets—white foam coffee cups for example—it is EPS. XPS is fine celled foam. Both EPS and XPS have low melting points and can be easily cut in the field. XPS requires the use of control joints and mechanical fasteners. It's often cut into architectural trim and other shapes, such as those shown in *Figure 19*.

NOTE

Masonry and concrete structures with no exposed wood do not require a water barrier.

Figure 19 Polystyrene architectural trim.
Source: Allison H. Smith/Shutterstock

In EIFS installation, EPS insulation panels are joined together. A rasp is used to create a uniform, flat surface. The panels are backwrapped with a fiberglass mesh (*Figure 20*). Panels adjacent to door or window jambs must be backwrapped and then caulked with the recommended sealant to prevent leaks (*Figure 21*). The base coat is then applied. It should be primed to ensure a good finish coat.

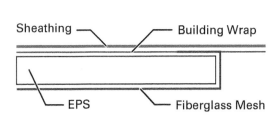

Figure 20 Backwrapping.

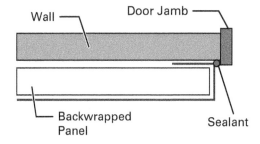

Figure 21 Backwrapped panel.

Lastly the synthetic stucco elastomeric finish coat is applied. When finishing EIFS, it is important to keep a wet edge as you work. Additionally, when installing large areas, it's good practice to create joints within the finished surface.

This provides separate blocks that allow for stopping points during installation, and it helps with maintenance as you only need to refinish one block.

Once installed, it is often difficult for even an experienced inspector to distinguish stucco from EIFS. One telltale sign is that EIFS sound hollow when tapped, but stucco sounds solid.

When properly installed, EIFS can provide years of good appearance and service. EIFS durability relies on its ability to maintain a watertight seal over the exterior framing. Any failure in the wrapping can result in serious water damage. Most EIFS failures are due to improper installation of the wrap, the flashings, or the system itself. One type of EIFS failure that is unrelated to installation occurs when the framing dries out and the building moves, creating openings in the system.

3.4.0 Fiber-Cement Siding

Fiber-cement siding is manufactured using portland cement, sand, fiberglass and/or cellulose fiber, selected additives, and water. It is usually pressure formed and heat cured. This material can be molded into a variety of shapes, including panels (*Figure 22*).

Figure 22 Fiber-cement siding.
Source: David Papazian/Shutterstock

Fiber-cement siding has some major advantages over hardboard or wood siding: it will not rot, it is noncombustible, and it can withstand a termite attack. Fiber-cement siding is highly resistant to impact damage and, in some cases, can withstand hurricane-force winds of 130 mph or more. It also resists permanent damage from water and salt spray. This siding is especially well suited for use in areas prone to fires or high winds.

Woodgrain Shingle Siding

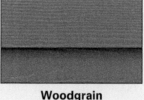

Woodgrain Lap Siding

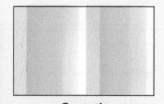

Smooth Vertical Siding

Source: © 2024 James Hardie Building Products Inc. All Rights Reserved.™

As with wood siding, fiber-cement siding is available in single-lap siding in widths from 6 to 12 inches, and as vertical panels. Commercial buildings are more likely to use the panels as cladding because they can be installed quickly and have fewer joints. The panels come in a number of different surface patterns. The recommended finish is 100 percent acrylic latex paint over an alkali-resistant primer. However, gloss or satin oil/alkyd paints over an alkali-resistant primer may also be used. The panels cannot be stained.

To achieve satisfactory performance, and for warranty purposes, the siding must be installed and finished as specified by the manufacturer. The following are general guidelines for installing fiber-cement siding:

- The siding may be applied over walls sheathed with wood or insulation board up to 1" thick and with studs spaced not more than 24" OC. The lowest edge of the siding should not be in contact with the earth or standing water. When cutting the siding, be sure to prime all cut ends with an alkali-resistant primer.

- Store the panels flat, keeping them dry and covered while in storage. Avoid installing wet panels. Wet panels swell, and as they dry, they shrink. Due to this shrinkage, the joints of panels that are installed wet will pull apart.

- To avoid bending the panels, carry them upright and on edge. Use a two-person carry as necessary. The panels are durable, but the corners and edges are prone to breakage.

- A weather-resistant barrier, such as Tyvek® or HardieWrap®, may be required under the siding when it is applied directly to studs or over wood sheathing. Consult the manufacturer's instructions for specific requirements.

- Fiber-cement siding may be cut using a circular saw with a variety of specialized fiber-cement blades. Usually, these blades are made of carbide and are tipped with polycrystalline diamond powder (*Figure 23*).

WARNING!

Because this dry material will be drilled, cut, and/or abraded, proper respiratory protection must always be used to avoid inhaling toxic silica dust that can cause a fatal lung disease called silicosis.

Figure 23 Polycrystalline diamond–tipped saw blade.
Source: Courtesy Milwaukee Tool

- Only galvanized steel, copper, or stainless steel flashing and screws/nails may be used when installing fiber-cement siding. The size and style of fasteners used on a job depends on the height of the building, the maximum wind speed in the area, and the use of the building. Consult with the manufacturer of the siding to determine the correct fasteners for the job. Never use staples.

- Galvanized steel with a powder or baked enamel finish or vinyl inside/outside corners and other trim can be used and painted to match the siding finish. Never use aluminum trim components or fasteners; they will corrode when in contact with galvanized enamel-coated steel siding.

- At horizontal joints, maintain a $\frac{1}{4}$" space between panels, and install Z-flashing.

- Maintain a minimum of 1" to 2" between the bottom of the panel and decking, steps, driveways, or other surfaces, excluding the finished grade. Maintain a minimum of 6 inches between the bottom of the panel and the finished grade.

- Always prime the product as specified by the manufacturer.

- Always install the product according to the manufacturer's instructions.

3.5.0 Glass Fiber Reinforced Concrete (GFRC)

Glass fiber reinforced concrete (GFRC) is a concrete product reinforced with glass fibers to make it durable and lightweight. GFRC can be made into panels or molded into various architectural shapes. It can even be produced to match sections of older buildings that have been damaged.

GFRC products can be installed over wood frame or concrete/masonry construction. When installed over wood, a moisture barrier must be applied to the structure. GFRC panels can be installed with or without insulation. Although the panels are made of concrete, they are not strong enough to be load-bearing. Because GFRC products are fabricated, they are consistent in size and color. They are available in a number of finishes, including aggregate and brick.

GFRC products may be installed with fasteners through the front of the panel or with hanger devices on the rear of the panel. All fasteners must be corrosion resistant and should match the panel color if they penetrate the front side. Panels should be installed according to the manufacturer's instructions. Some producers specify the need for control joints in certain installations; others may specify joint gaps between panels. All gaps should be filled with the recommended type of sealant.

3.0.0 Section Review

1. True or False? Traditional stucco is breathable and allows moisture to pass through whereas acrylic stucco is waterproof.
   a. True
   b. False

2. Once all synthetic stones have been set in place, the spaces between them are filled with _____.
   a. cement
   b. grout
   c. sand
   d. crushed rock

3. Exterior insulation and finish systems (or EIFS) combine insulation with a synthetic, textured _____.
   a. mortar and grout
   b. paint finish
   c. fine celled foam
   d. exterior coating

4. The major advantages fiber-cement siding has over hardboard or wood siding are that it will not rot, is noncombustible, and can withstand _____.
   a. hurricane winds
   b. high water levels
   c. a termite attack
   d. abrasion

5. Glass fiber reinforced concrete (GFRC) is a concrete product reinforced with glass fibers to make it durable and _____.
   a. soft
   b. lightweight
   c. waterproof
   d. thick

Module 45205 Review Questions

1. The primary purpose of exterior cladding is to _____.
   a. hide blemishes
   b. protect the structure
   c. look good
   d. provide insulation

2. When wrapping a building, the beginning of a new roll should overlap the end of the previous roll by _____.
   a. 1" to 6"
   b. 6" to 12"
   c. 12" to 18"
   d. 18" to 24"

3. What is the purpose of a control joint?
   a. Drain water from behind the cladding
   b. Cover vertical seams to prevent water infiltration
   c. Support the insulated panels of EIFS
   d. Relieve coating stress due to thermal changes

4. What is the purpose of a weep screed?
   a. Drain water from behind the cladding
   b. Cover vertical seams to prevent water infiltration
   c. Support the insulated panels of EIFS
   d. Relieve coating stress due to thermal changes

5. What is the use for a starter strip?
   a. Cover vertical seams to prevent water infiltration
   b. Support the insulated panels of EIFS
   c. Relieve coating stress due to thermal changes
   d. Prevent water from seeping behind the cladding

6. What is the use for flashing?
   a. Cover vertical seams to prevent water infiltration
   b. Support the insulated panels of EIFS
   c. Relieve coating stress due to thermal changes
   d. Prevent water from seeping behind the cladding

7. Traditional stucco is _____.
   a. waterproof
   b. cement-based
   c. obsolete
   d. crack resistant

8. Stucco cladding over a concrete structure is applied over _____.
   a. building wrap
   b. metal mesh
   c. scratch and brown coats
   d. chicken wire

9. True or False? Stucco finishes tend to expand and contract with temperature changes.
   a. True
   b. False

10. Expansion joints in stucco are placed _____.
    a. every 100 feet
    b. at lath joints
    c. below grade
    d. wherever it is convenient

11. A scratch coat applied over lath should be approximately _____ thick.
    a. $1/8$" to $1/4$"
    b. $1/4$" to $1/2$"
    c. $1/2$" to $3/4$"
    d. $3/4$" to 1"

12. A brown coat is a mixture of _____.
    a. browning and cement
    b. browning and sand
    c. sand and cement
    d. sand, browning, and cement

13. When cutting synthetic stone, be sure to _____.
    a. wet the stone before you cut it
    b. grind off any rough edges
    c. clean the stone after it is cut
    d. grout the stone before you cut it

14. When placing cut stones up to an inside corner, place them _____.
    a. so the grout lines are straight
    b. in a long-short pattern
    c. cut side facing out
    d. $1/2$" from the wall

15. To keep flat stones level when installing them, mark a line every _____ as a guide.
    a. 4" to 8"
    b. 6" to 10"
    c. 8" to 12"
    d. 10" to 14"

16. If you spatter grout on a synthetic stone after it is placed, you
    should _____.
    a. wipe the grout away with a damp cloth
    b. wait for the grout to dry, and then brush it away
    c. remove the stone from the wall, and soak it in water
    d. realize that grout spatters cannot be removed from synthetic stones

17. When setting synthetic stone, use a striking tool to _____.
    a. fit the stones
    b. firmly set the stones
    c. finish the grout and remove any excess
    d. scratch the brown coat

18. Extruded polystyrene is _____.
    a. made of pellets of polystyrene molded into a shape
    b. a fine celled foam
    c. mistakenly called Styrofoam™
    d. used in coffee cups

19. One reason for EIFS failure in wooden framed buildings is that _____.
    a. the wood dries, causing the building to move
    b. the buildings are susceptible to termite damage
    c. wood swells in humid weather
    d. polystyrene breaks down quickly

20. Fiber-cement siding should be finished with _____.
    a. latex paint
    b. varnish
    c. stain
    d. cement sealer

21. An acceptable way to fasten fiber-cement siding to a stud is to use _____.
    a. staples
    b. plastic fasteners
    c. galvanized steel screws
    d. aluminum nails

22. At horizontal joints, install fiber-cement panels _____ apart.
    a. $\frac{1}{8}$"
    b. $\frac{1}{4}$"
    c. $\frac{3}{8}$"
    d. $\frac{1}{2}$"

23. Fiber-cement siding should be installed _____ above decking.
    a. $\frac{1}{2}$"
    b. $\frac{1}{2}$" to 1"
    c. 1" to 2"
    d. 2"

24. True or False? GFRC can be installed over insulation.
    a. True
    b. False

Answers to odd-numbered questions are found in the Module Review Answer
Key at the back of this book.

Answers to Section Review Questions

Answer	Section	Objective
Section One		
1. c	1.1.0	1a
2. a	1.2.0	1b
3. a	1.3.0	1c
Section Two		
1. b	2.1.1	2a
2. d	2.1.2	2b
3. b	2.1.3	2c
4. a	2.1.4	2d
Section Three		
1. a	3.1.0	3a
2. b	3.2.0	3b
3. d	3.3.0	3c
4. c	3.4.0	3d
5. b	3.5.0	3e

MODULE 45206

Interior Finishes

Objectives

Successful completion of this module prepares you to do the following:

1. Describe the various types of interior paint.
   a. Describe the purpose of pigments, binders, solvents, and additives in interior paint.
   b. Identify the surfaces for which the various types of interior paint are best suited.

2. Identify the tools commonly used by professionals to prepare a drywall surface, mix materials, and apply materials to a drywall surface.
   a. Identify the tools used to prepare drywall for painting.
   b. Identify the tools used to prepare interior paint and texturing compound for application.
   c. Identify the tools used to apply paint and texturing compound to drywall surfaces.

3. Describe the fundamental methods used in surface preparation and application of interior paint.
   a. Describe the steps that should be taken to ensure a drywall surface is prepared to receive paint.
   b. Describe the key considerations and steps involved in mixing paint and texturing compound.
   c. Describe the key considerations and steps involved in applying primer to interior surfaces.
   d. Describe the key considerations and steps involved in spray painting interior surfaces.
   e. Describe the techniques used to apply interior paint with hand tools.

4. Identify the different types of interior specialty finishes.
   a. Describe specialty texture finishes and the techniques used to apply them to interior surfaces.
   b. Describe specialty faux finishes and the techniques used to apply them to interior surfaces.

Performance Tasks

Under supervision, you should be able to do the following:

1. Use a mixer to prepare paint and compound for application.
2. Set up and use a paint sprayer to apply an even coat of paint to an interior surface.
3. Set up and use, or explain the use of, a portable and/or a self-contained texturing machine to apply texture to an interior surface.
4. Use brushes, rollers, and other hand tools to cut-in around fixtures and other tight areas.
5. Determine the surface area of an interior surface and prepare an appropriate amount of paint.
6. Apply, or explain the application of, specialty texture and faux finishes to an interior surface to achieve a desired appearance.

Digital Resources for Drywall

Scan this code using the camera on your phone or mobile device to view the digital resources related to this craft.

Overview

Once the walls have been framed and the drywall is in place, it's time to apply the final touches. Painting is an affordable and efficient way to beautify the interior of a structure. Although tile, wood, paneling, and wallpaper are sometimes used to accent a surface, painting is the preferred method to finish the majority of interior walls in both commercial and residential buildings.

This module introduces you to various types and characteristics of interior paint, the tools used to prepare and apply paint, as well as basic and specialty painting techniques. When performing finishing work, remember that a good finish job is a dynamic process that starts with professional prep work and continues until the final coat is applied to a surface.

NCCER   Industry-Recognized Credentials

If you are training through an NCCER-accredited sponsor, you may be eligible for credentials from NCCER. The ID number for this module is 45206. Note that this module may have been used in other NCCER curricula and may apply to other level completions. Contact NCCER at 1.888.622.3720 or go to **www.nccer.org** for more information.

You can also show off your industry-recognized credentials online with NCCER's digital credentials. Transform your knowledge, skills, and achievements into credentials that you can share across social media platforms, send to your network, and add to your resume. For more information, visit **www.nccer.org**.

1.0.0 Introduction to Interior Paint Products

Performance Tasks

There are no Peformance Tasks in this section.

Objective

Describe the various types of interior paint.

a. Describe the purpose of pigments, binders, solvents, and additives in interior paint.

b. Identify the surfaces for which the various types of interior paint are best suited.

Paint can drastically affect the mood of a space and spur heated debates among those involved in making the color choice. Fortunately, in commercial construction, painters are not responsible for decisions about color. In addition to color, other characteristics of paint have a more subtle, but critical, impact on visual appeal and durability. The sheen, usually referred to as *finish,* of paint is a key consideration because it affects the ease of application and cleaning as well as longevity. The base of the paint (oil, latex, acrylic, or a combination) is also an essential factor to consider when selecting paint. Another aspect to keep in mind are additives that are used by manufacturers to enhance the performance of paint.

1.1.0   Pigments, Binders, Solvents, and Additives

Paint is a precise mixture of several components, including pigments, binders, solvents, and additives.

1.1.1 Pigments

Pigments give paint its color. Paint pigments may be naturally occurring materials as well as synthetic substances derived from natural pigments. Pigments absorb some wavelengths of light and reflect others, which results in the visual effect we know as *color perception. Figure 1* illustrates this phenomenon. White light—the natural light around us that contains all colors of the spectrum—shines on an object, but due to the pigments in the object only some of that light is reflected off the object, resulting in the colors we see.

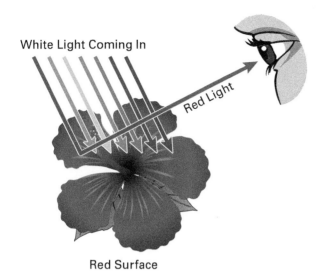

White Light Coming In

Red Light

Red Surface

Figure 1 How color is perceived.

Paint is available as a pre-mixed color, or it can be custom colored just before purchase. Retailers use computerized paint mixers that store formulas that match their color samples. These machines add precise amounts of liquid pigments to the mixer to achieve the selected color. Some vendors duplicate colors by using computers to analyze a sample color and determine the formula needed to reproduce the color. Desired shades are achieved by mixing different colored pigments into a tint coat, which is usually white, off-white, or light beige.

1.1.2 Binders

Also called *resins*, binders make the paint dry into a film. As the name suggests, binders provide the cohesiveness necessary to hold the pigments and other components together. The two main categories of binders used in interior paint are water-based and oil-based. Water-based binders are made from synthetic, water-soluble polymers of acrylic acid. Prior to the 1940s, water-based paints were made with actual latex—a rubbery substance that comes from a tree. Today, synthetic latex products (including acrylics, polyvinyl acetate, and styrene butadiene) are used instead of natural latex. Despite the fact that water-based paints do not contain any actual latex, they are still commonly referred to as latex paints.

Oil-based paints use natural or synthetic resins derived from oil as their binders. Linseed oil, derived from the flax plant, is the most common natural oil used as a binder. Other natural oils used to bind paint include tung, poppy, walnut, and safflower. Synthetic oils, such as alkyds and polyurethane, are made from petroleum.

1.1.3 Solvents

Solvents dissolve the binder and make it liquid. They also hold the pigment in suspension, keeping it evenly distributed throughout the paint. When paint is applied to a surface, the solvent dries, leaving a film of binder and pigment. Binders and pigments are referred to as *paint solids*. Synthetic latex paint uses water as its primary solvent. It also uses other solvents, but in very small amounts. Oil-based or alkyd-based paints use solvents derived from petroleum, such as mineral spirits.

1.1.4 Additives

The paint industry spends a great deal of money researching how to improve its products. Additives used to enhance paint include the following:

- *Wetting agents* — help the paint to spread easily
- *Drying agents* — speed the paint's drying time
- *Bonding agents* — increase paint's ability to adhere to a surface
- *Antimicrobials* — control the growth of mold and mildew
- *Perfumes* — add temporary scent to paint to mask odors

NOTE

Many people are allergic to latex. Because modern latex paints contain a synthetic form of latex, they will not cause allergic reactions in these individuals.

CAUTION

Water-based paints allow water to pass through, even after drying, but oil-based paints do not. As a result, oil-based paints can crack and peel if applied over moist surfaces. In addition, water-based paints cannot be used to paint over oil-based paints because the two paints will not adhere.

1.2.0   Types and Uses of Interior Paint

There are a variety of interior paint types and finishes. Whether you need a water-based latex paint to touch up a wall or an oil-based paint to cover new interior trim, different types of paints are available for practically every need. This section covers the most common types of interior paint and finishes.

1.2.1 Water-Based

Acrylic latex: A type of paint that is water-based with an acrylic binder. Acrylic latex paints have the enduring color, cleanability, and adhesion of latex with the added durability of acrylic resin.

Of the various water-based paints, latex is the most commonly used on interior walls. **Acrylic latex**, a combination of synthetic latex and acrylic resin, is the predominant variety. This paint dries quickly, has low odor and fewer volatile organic compounds (VOCs), has long color life, is washable with water, and resists peeling and cracking. The addition of the acrylic binder improves adhesion and durability compared to latex paint without acrylic resin. Acrylic latex paints are very safe to use, but you should always check the product literature and SDS (Safety Data Sheet) for the personal protective equipment you must wear. One common adverse reaction some people have to latex paint is contact dermatitis (inflammation of the skin).

1.2.2 Oil-Based

Oil-based paints are normally not used to cover large, interior surfaces. They are not as easy to work with as water-based paints and cannot be cleaned with water. In addition, they emit more vapors into the air, take longer to dry, and tend to yellow over time. Oil-based paints, however, are well-suited for high-contact areas like trimwork, hallways, and cabinets because of their hardness, durability, and resistance to stains.

Vitreous enamel, sometimes referred to as *porcelain enamel*, is made by melting hard, glasslike minerals together. Enamel paint is an oil-based paint containing vitreous enamel, which causes it to dry to a glossy, hard finish. This type of paint is very durable and scratch resistant.

1.2.3 Textured

Textured paints are an easy way to add interest to a room. Sometimes surfaces are textured with a compound similar to joint compound before they are painted. This method is described later in this module. Another, simpler method is to add texturing materials to paint before it is applied to a surface. Some common texturing materials include sand, clay, vermiculite, and polystyrene. Depending on the desired effect, the texture of these materials may be fine, coarse, or anything in between. Texturing a surface is an easy way to conceal imperfections.

1.2.4 Epoxies

Epoxy is a substance formed from chemicals called epoxides that undergo a chemical reaction and become extremely hard. Epoxies may be applied separately as a top coat over a painted surface or they may be mixed into paint before application. Epoxy paints are often used in hospitals, schools, and restaurants, where surfaces must be durable, resistant to chemicals, and washable. Epoxy paints are typically used on block or concrete, but they are also formulated for drywall. Some epoxy paints are single-component paints, while

> **Did You Know?**
>
> **Ceiling Paint Is a Little Different**
>
> Gravity can be a challenge when painting a ceiling. To keep paint from dripping down, ceiling paint is more viscous—thicker and stickier. Another issue with ceilings is that they are typically white. This is good for brightening the space and complementing other colors in the room, but it makes it difficult to distinguish a fresh coat of white paint from the existing paint or drywall. To solve this problem, ceiling paint is often treated with an additive to make it a slightly blue or pink color while it is wet. Many ceiling paints are also formulated to minimize glare and to retain the acoustical properties of the ceiling.

others have two components. Single-component epoxy paints have the convenience of being premixed and, when stored properly, they have a long shelf life. Two-component epoxy paints must be thoroughly mixed before use. They have a **potlife** of only a few hours.

Potlife: The period of time in which a mixed, two-component epoxy paint must be used.

Volatile Organic Compounds

Volatile organic compounds (VOCs) are gases that evaporate quickly and are easily inhaled or absorbed through skin or eyes, making them potential health hazards (*Figure 2*). Certain paints, solvents (such as mineral spirits), varnishes, and waxes contain VOCs. You can take the following steps to protect yourself:

- Use low VOC products when available.
- Make sure the area you are working in is well ventilated.
- Use a respirator if necessary.
- Use a VOC monitor to measure the amount of VOCs in your work area.
- Read the warning labels on the product for information about hazardous vapors.
- Check the product's safety data sheet and use the recommended PPE.

Short-term exposure to VOCs can cause eye, nose, and throat irritation, headaches, loss of coordination, and nausea. If you experience any of these symptoms, leave the work area and seek medical assistance. Long-term exposure can damage the liver, kidneys, and central nervous system.

In addition to being health hazards, VOCs contribute to atmospheric pollution and are regulated by the US Environmental Protection Agency (EPA) and many state agencies.

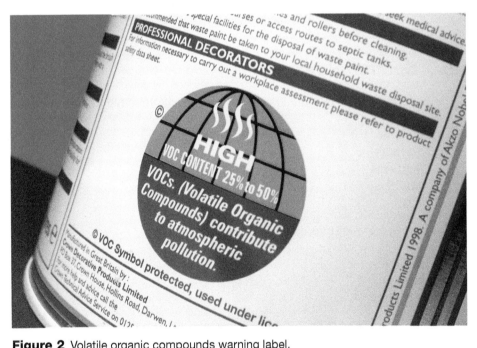

Figure 2 Volatile organic compounds warning label.
Source: Peter Alvey/Alamy Images

1.2.5 Paint Sheen (Finishes)

Sheen, often referred to as *finish,* refers to the shininess of the paint when applied to a surface and dried. Sheen is a result of the relationship between the amount of pigment and the amount of binder in the paint. The combination of more pigment and less binder reduces the amount of sheen and vice versa. This relationship is called the **pigment-volume concentration (PVC)**. Highly pigmented paints provide better coverage but are less durable due to the lower levels of binder. Paints with lower PVC levels—or higher levels of binder compared to pigment—are not as good at covering but they are more durable, cleanable, and moisture resistant (*Table 1*).

Pigment-volume concentration (PVC): The relationship between the amount of pigment and the amount of binder in paint. Higher PVC levels indicate more pigment compared to binder.

TABLE 1 Varieties of Paint Sheen

Sheen/Finish Type	Pigment-Volume Concentration (PVC)	Description
Flat/matte	40% or higher (highest level of pigment; lowest amount of binder)	Absorbs light, no shine, good for hiding imperfections, easy to apply, difficult to clean, generally more affordable. Used in low traffic/contact areas.
Eggshell	35–40%	Reflects a small amount of light, slight amount of sheen. Used in areas with moderate traffic/contact.
Satin	30–35%	Average amount of light reflectance (sheen), mildly moisture resistant, suitable for high traffic/contact areas, good washability.
Semigloss	25%	Above average light reflectance (noticeable shine), high degree of moisture resistance, durable, easy to clean. Good for high traffic/contact areas. Used to accent focal areas.
High gloss	15% (lowest amount of pigment; highest amount of binder)	Very shiny (high light reflectance), high moisture resistance, very durable, excellent washability. Good for high traffic/contact areas. Used to accentuate decorative features (cabinets and trim) and to enhance the natural warmth of a space.

1.0.0 Section Review

1. Which component of paint absorbs some wavelengths of light and reflects others?
   a. Binders
   b. Solvents
   c. Pigments
   d. Additives

2. Wetting agents, drying agents, bonding agents, antimicrobials, and perfumes are examples of paint _____.
   a. solvents
   b. additives
   c. binders
   d. pigments

3. Which of the following is a combination of synthetic latex and acrylic resin and is also the predominant variety of paint used on interior walls?
   a. Latex
   b. Epoxy
   c. Acrylic
   d. Acrylic latex

4. The variety of paint with the highest pigment volume concentration (PVC) is _____.
   a. high gloss
   b. satin
   c. eggshell
   d. flat/matte

2.0.0 Interior Painting Tools

Objective

Identify the tools commonly used by professionals to prepare a drywall surface, mix materials, and apply materials to a drywall surface.

   a. Identify the tools used to prepare drywall for painting.

   b. Identify the tools used to prepare interior paint and texturing compound for application.

   c. Identify the tools used to apply paint and texturing compound to drywall surfaces.

Performance Tasks

There are no Performance Tasks in this section.

Painting and texturing drywall requires specialized tools for preparing the surface, mixing the paint or texturing compound, and applying the paint or compound to create the desired effect. The surface area to be finished, materials being mixed, textures added to the paint, and degree of detail work are factors that are considered when selecting the right tools for the job.

2.1.0 Tools for Surface Prep

Most surface preparation should be completed during the installation of the drywall. However, the drywall surface should always be inspected prior to painting so that any defects, blemishes, or uneven areas can be corrected. Additionally, the surface may need to be cleaned to remove dust left behind from the initial installation and finishing process. Any items on the surface that will not receive paint should be covered or temporarily removed. Final surface preparation, or reworking, may require the following tools:

- Screwdriver or screw gun to take down items
- Drywall knives (also called *tape knives* or *joint knives*) to reapply compound
- Drywall compound (usually called *mud*)
- Mud pan to hold drywall mud for application
- Sanding block, electric sander, or sponge sander for sanding small areas of dried compound
- Vacuum, sponge, cloths, or mopping tool

NOTE

A comprehensive explanation of tools for finishing drywall is included in NCCER Module 45105, *Drywall Finishing*. The tools mentioned here are those needed to make minor surface corrections to drywall before painting.

2.2.0 Tools for Mixing

Paint and compound mixers are used to mix paint, and to mix water and texturing materials into powder compounds. Some mixers are specially designed to mix compounds, paints, or both. Mixers are either self-contained models or separate attachments that are used with power drills or mixer drills. The self-contained mixer shown in *Figure 3* is programmable with different settings for mixing mortar, grout, plaster, texture compounds, and paint. *Figure 4* shows a self-contained, hand-held, dual-handle paddle mixer with variable speeds for mixing different materials. Dual-handled mixers and specialized mixing drills are preferrable when mixing large quantities of material or for prolonged mixing tasks.

    Mixing paddles come in a variety of shapes and sizes. A standard power drill with a chuck (wheel and axle-like mechanism that holds the drill bit) can be used with mixing paddles that have a hexagonal shaped shaft. The diameter of the paddle's shaft determines the size of the chuck needed to secure the paddle in the drill. Specialized mixing drills normally require a specific paddle with a shaft that is threaded to fit the drill.

CAUTION

Standard cordless power drills should only be used to mix small amounts of nonviscous materials. Standard drills are designed to rotate at high speeds, but they are not designed to generate the amount of torque needed to mix thick materials. Furthermore, running a standard drill at maximum capacity for long periods can cause it to overheat. For dense materials like textured compound, use a dedicated or self-contained mixing drill.

Figure 3 Power mixer with programmable settings.
Source: RIDGID Tool Company

Figure 4 Self-contained, dual-handle mixer.
Source: Courtesy of Stanley Black & Decker, Inc.

2.3.0   Tools for Applying Paint or Texture

Achieving the desired finish requires knowledge of application techniques and tools. Some surfaces can be covered quickly with a paint sprayer, while others may require the use of manual tools like rollers and brushes. If a texture is being applied to the surface, it may be done with a texturing machine or by hand. Each method of application has a unique set of tools with which a finishing technician must be familiar.

2.3.1 Paint Sprayers

With paint sprayers, large areas can be painted relatively quickly and easily. Sprayers also simplify the task of painting corners and ceilings. There are three basic types of paint sprayers (*Table 2*): airless, pneumatic (*Figure 5*), and HVLP (high velocity low pressure *Figure 6*).

TABLE 2 Types of Paint Sprayers

Sprayer	Features
Airless	Uses a powered pump (usually corded but can be gas or battery)
	Does not require thinning of paint
	Paint is not mixed with air
	Covers large surface area quickly
	Preferred method for commercial jobs
Pneumatic	Uses compressed air
	Typically has attached canister for paint
	Used on small areas
	Even coverage
	Lots of overspray
	Also referred to as *siphon-fed spray gun*
High velocity low pressure (HVLP)	Uses compressed air but at low pressure
	Less overspray
	Some can pull paint directly from a bucket
	Good for small to medium projects
	Also referred to as *gravity-fed spray gun*

Figure 5 Pneumatic spray gun.
Source: Courtesy of Stanley Black & Decker, Inc.

Figure 6 High velocity low pressure (HVLP) spray gun.
Source: Courtesy of Stanley Black & Decker, Inc.

Paint sprayers vary by manufacturer. Always be sure to study the operator's manual before using an unfamiliar piece of equipment. The following are general guidelines for using sprayers:

- Before starting any paint job, make sure that the equipment is clean. Check and clean all filters. Wipe dust and dirt from the equipment to avoid contaminating the paint.
- Disconnect the power source before performing any servicing of the equipment. To release any built-up pressure, point the gun into the reservoir and pull the trigger.
- Check the manufacturer's instructions to determine the proper viscosity of the paint.
- Practice spray painting on scraps of drywall. Before you start painting, adjust the sprayer pressure and gun aperture until the paint flow is correct.
- Hold the gun so that it is 8" to 12" from the surface being painted.
- Position the gun at a 90-degree angle to the surface being painted to prevent painting in an arc pattern.
- Move the gun at a rate of speed that will allow for adequate paint coverage. Begin moving the gun before pulling the trigger to prevent paint buildup.
- Overlap strokes to blend the paint.
- Frequently wipe excess paint from the tip of the gun to prevent clogging.
- Some textured paints cannot be applied with a sprayer. The texturing material will clog the sprayer. When applying textured paints, always check the paint manufacturer's recommendations before beginning the job.
- Do not leave paint in the reservoir of the sprayer for prolonged periods.
- Clean all sprayer equipment after use according to the manufacturer's specifications. This usually involves running the correct paint solvent through the sprayer until it is clean. Water is the appropriate solvent for latex paint. Always be sure to clean sprayer equipment properly before storage.
- Never point a spray gun at anyone, and never spray another worker with paint.

2.3.2 Spray Texturing Machines

Spray texturing machines are used to apply a texture to a surface before the surface is painted. The texturing material is placed in the hopper, which has a small hole at the bottom where the hopper joins the gun. This opening allows material to be fed into the gun. When the trigger is pulled, compressed air rushes past the hopper opening, and texturing material and air are shot out of the gun. The air rushing past the hopper opening also creates a suction that pulls the material into the gun.

Some texturing machines utilize a hopper mounted on a cart (*Figure 7*). With another type of texturing machine called a hopper gun, the operator holds a hand-held hopper with one hand while controlling the sprayer with the other.

Figure 7 Cart-mounted texturing machine.
Source: MARSHALLTOWN

Orifice: The opening in the nozzle of a paint sprayer machine through which paint is sprayed.

Another type of portable texture sprayer is shown in *Figure 8*. The texturing material is placed in a canister that is carried on the operator's back. This canister, which holds $2\frac{1}{2}$ gallons of texture, is fed compressed air from an external air compressor. The texturing material is applied with a gun. This design is easier to manage than the hopper gun arrangement because the operator is free from the need to support the hopper while spraying.

Large jobs require large equipment. Some of the largest spray texturing machines can hold up to 500 gallons of texturing compound.

Texturing guns need to have an adjustable **orifice** for applying small, fine textures as well as large, coarse textures. Using an orifice that is too small will clog the gun. Using one that is too large will apply too much texturing material. For ceiling applications, it is best to use an angled orifice. *Figure 9* shows a set of interchangeable orifices that attach to the nozzle of a paint sprayer.

Figure 8 Back-mounted texturing machine.
Source: MARSHALLTOWN

Figure 9 Orifice set for paint sprayer nozzle.
Source: MARSHALLTOWN

As with paint sprayers, texturing guns vary by manufacturer. Always be sure to study the operator's manual before using any equipment for the first time. The following are general guidelines for using a texturing machine:

- After using any texturing machine, be sure to clean it according to the manufacturer's specifications. Never store equipment until it has been properly cleaned.
- Before beginning any job, make sure that the equipment is clean. Wipe any foreign matter from the hopper to avoid contaminating surfaces. Substances such as oil and grease can prevent paint from bonding to the texturing compound.
- Prior to servicing any equipment, always disconnect the power source.
- Before beginning a job, practice applying the texturing material on scrap drywall.

- Hold the gun so that it is 12" to 16" from the surface being textured.
- The technique used when applying texturing materials depends on the desired result. Once a method has been selected, it must be maintained until the job is finished.
- When texturing a ceiling, use a gun that has an angled nozzle, if possible.

2.3.3 Hand Tools

Brushes and rollers are impractical for painting commercial areas, although they may find limited use for doing **cut-in** work around fixtures and touching up small areas. Because of their size, commercial spaces are usually spray painted. This allows the job to be completed quickly and the paint to be applied evenly. Even when applying a faux (simulated) finish to a surface, the base coat is usually applied with a sprayer. Hand tools, however, are used to apply the detail coats.

Cut-in: A technique used to manually paint around a fixture with a brush or roller.

Brushes

Cutting-in is the process of painting the edges of a surface that are adjacent to surfaces that are not intended to receive paint. This can be accomplished in a variety of ways depending on the amount of edge work and degree of difficulty, among other factors. Areas not to be painted can be taped off with painter's tape, but this method is very time consuming. Most pros become highly skilled at cutting-in with an angled paint brush so that taping and tools like straightedges are not needed.

Choosing the right brush for cutting-in is very important. Even the most skilled painter will struggle to create a clean, straight edge with the wrong brush. A 2" to 3" angled brush is ideal for most edging tasks because it is wide enough to paint away from the protected surface but small enough to maneuver easily. The shape of angled brushes makes them well-suited for fitting into tight spaces and being turned away from the surface to be avoided. In addition, angled brushes can be held in a way that allows the bristles to be dragged in a smooth, controlled manner across a surface.

Specialized texturing brushes are sometimes used for applying custom finishes. These specialized tools include stipple brushes (*Figure 10*), and single and double texture brushes, also known as crow's foot brushes (*Figure 11*).

CAUTION

Falls are one of the leading causes of injuries and fatalities in construction. Injuries, and worse, occur even when falling short distances from ladders, scaffolding, and mobile working platforms. Always use proper fall prevention and protection and follow all ladder safety rules.

NOTE

Paint brushes with synthetic bristles should be used when working with water-based paints. Natural bristles will absorb some of the water in the paint and become saturated. When painting with water-based paint, water can be used to clean brushes. However, a solvent is needed to clean brushes used with oil-based paint.

Figure 10 Stipple texture brush.
Source: MARSHALLTOWN

Figure 11 Double texture (crow's foot) brush.
Source: MARSHALLTOWN

Rollers

The speed and efficiency of paint sprayers make them the primary tool for finishing large surface areas quickly. However, rollers are used on occasion to *backroll*, or paint over a surface again while it is still wet. Backrolling is done to get rid of streaks, runs, or areas of uneven coverage created by spraying. Rollers are also used at times to intentionally create a subtle texture in the paint that cannot be achieved with a sprayer. The rolling method of painting is much more time consuming than spraying, but it provides thicker and more even coverage, which is why it is still used in some situations.

The size of the roller, the thickness (nap) and composition of the roller cover, the surface to be painted, and the type of paint being applied are factors to consider when selecting materials for rolling paint. General tips for selecting rollers include the following:

- Use roller covers with synthetic fibers for water-based paints.
- Covers with natural fibers are better for oil-based paints.
- Use covers with a shorter nap ($\frac{1}{4}$"–$\frac{3}{8}$") for smooth surfaces like drywall.
- Use a thicker nap for rough or porous surfaces.

To eliminate the need for ladders and scaffolding, use a telescoping pole to reach high areas with a roller.

Another use for rollers is creating patterned textures that resemble stucco, leather, or marble, among others. Because rollers are customizable, they may also be embossed with just about any design one can imagine, including floral motifs, geometric patterns, ocean waves, and wood grain. The rollers in *Figure 12* are used to create stippling and stucco as well as other abstract effects.

There are numerous hand tools that a drywall technician may need depending on the type of finish being applied. *Table 3* describes the common, and some of the less common, manual tools that are utilized to create flat, faux, and textured finishes with paint or drywall compound.

Figure 12 Texture rollers.
Source: Courtesy of Bon Tool Co.

TABLE 3 Manual Tools

Tool	Description	Purpose	Uses
Stucco brush	Wide, stiff-bristled brush with a handle.	Creates a variety of textures from stippling to swirls.	Texturing compound
Texture brushes	Stiff-bristled brushes with or without a handle. May be tandem-mounted to cover a larger area.	Create a variety of textured patterns.	Texturing compound
Texture rollers	Available in a variety of materials and textures depending on the finish desired. Rollers slip onto a holder that has handles. The handle can be attached to an extension for reaching overhead.	Create a variety of textures.	Texturing compound
Wipe-down blade	A wide, hardened steel blade with rounded corners to prevent gouging. May have a long or short handle.	Wipes excess material from a surface. Helps clean walls and floors after application of texturing materials.	Texturing compound
Paint roller and pan	Available in a variety of widths and materials to suit the type of paint used and the finish desired. Roller covers slip onto a handled frame. The handle can be attached to an extension for reaching overhead. Some rollers have a pump-fed option to continuously supply paint rather than using a pan.	Applies paint.	Smooth and textured paint
Flat-blade knife	Short-handled metal or plastic blade that comes in varying widths.	Applies texturing material; also used for troweled finishes.	Texturing compound
Sponges	Synthetic or natural (sea) sponge.	Apply free-form texturing finishes.	Texturing compound and faux finishes
Rags and paper	Clean cloths and soft or stiff paper.	Apply free-form texturing finishes.	Faux finishes
Whisk broom	Stiff bristled brush.	Creates a variety of textures from stippling to swirls. Similar to a stucco brush but can be used to produce a bolder brush pattern.	Texturing compound
Hawk and trowel	Flat plate with handle on bottom for holding material (hawk). Rectangular or triangular metal or plastic blade with a handle (trowel). Trowel edges may be smooth or have a pattern.	Apply texturing materials; also used to achieve troweled or knockdown finishes.	Texturing compound

2.0.0 Section Review

1. What is the first step that should be taken to prepare a drywall surface for painting?
   a. Vacuuming
   b. Sanding
   c. Inspecting
   d. Wiping down

2. A standard cordless power drill is appropriate for mixing _____.
   a. small amounts of nonviscous paint or compound
   b. textured compound
   c. thick paint or compound
   d. large amounts of paint or compound

3. The three basic types of paint sprayers are pneumatic, high velocity low pressure (HVLP), and _____.
   a. low velocity high pressure (LVHP)
   b. airless
   c. hydraulic
   d. self-contained

4. Hand tools, or manual tools, have a variety of uses including applying texturing compound, applying smooth and textured paint, and _____.
   a. creating knock-down finishes
   b. creating orange-peel textures
   c. covering large surface areas
   d. creating faux finishes

3.0.0 Surface Preparation, Mixing, and Application

Objective

Describe the fundamental methods used in surface preparation and application of interior paint.

a. Describe the steps that should be taken to ensure a drywall surface is prepared to receive paint.

b. Describe the key considerations and steps involved in mixing paint and texturing compound.

c. Describe the key considerations and steps involved in applying primer to interior surfaces.

d. Describe the key considerations and steps involved in spray painting interior surfaces.

e. Describe the techniques used to apply interior paint with hand tools.

Performance Tasks

1. Use a mixer to prepare paint and compound for application.

2. Set up and use a paint sprayer to apply an even coat of paint to an interior surface.

3. Set up and use, or explain the use of, a portable and/or a self-contained texturing machine to apply texture to an interior surface.

4. Use brushes, rollers, and other hand tools to cut-in around fixtures and other tight areas.

5. Determine the surface area of an interior surface and prepare an appropriate amount of paint.

A finish coat on walls and ceilings is one of the first things that people notice when they walk into a room. Whether the finish is smooth or textured, it should look professional, neat, and attractive. Applying the final finish to a surface is only one step in the entire process. Finish work begins with surface preparation.

3.1.0 Surface Preparation

A perfectly smooth or perfectly textured finish is not possible without a well-prepared surface. Depending on the desired end result, some cleaning, sanding, and refinishing may be necessary. Surfaces finished with texturing compound do not need to have smoothly sanded compound on every joint and fastener head. On the other hand, surfaces finished with smooth paint show every bump, depression, and ridge. This is especially true when the final finish is a dark-colored, high-gloss paint, which magnifies any defects.

Before you apply a finish to any surface, always carefully examine the surface for imperfections that will show through the final application. Drywall finishing procedures are covered in detail in NCCER Module 45105, *Drywall Finishing*. The following guidelines will help you find and correct minor surface imperfections before painting drywall:

- Use a systematic approach. Start at one corner and work your way around the entire area.

- Understand the degree of surface preparation that is required. Smooth finishes need to be applied to smooth surfaces. Sand any bumps, ridges, and high spots as needed.

- Examine all joints and fastener holes, and make sure that they are covered with enough joint compound and are adequately sanded. Flag any areas that need to be reworked. One way to flag an area is by using colored Post-it® notes. Another way is to use a marker or colored chalk to draw an arrow on the floor (provided the floor is bare).

- Check all corners to be certain that the corner bead is adequately covered.

- Look for any pits, gaps, depressions, shrinkage, or flaking in the joint compound, and repair as needed.

- Inspect the drywall surface for cracks, and repair as needed.

- When a *skim coat* is used, examine the entire surface, and make sure that all areas have been evenly coated, smoothed, and sanded.
- Cover any fixtures that will not be coated, such as windows and electrical fixtures.
- Thoroughly remove dust from surfaces to be painted. Use a vacuum with a soft-bristled, wide attachment followed by a sponge, damp cloth, or a large flat mopping tool with a slightly damp head.

WARNING!

Drywall compound contains silica dust and other substances linked to respiratory symptoms. Always wear an N95 respirator when sanding drywall compound or working around drywall dust. Vacuum dust from your clothing and wash your hands and arms thoroughly when finished.

3.2.0   Mixing Paint and Texturing Compound

When paint sits for long periods, the solids in it, including any texturing material, settle to the bottom of the container. Depending on the paint's formula, this process can take several weeks or longer. However, when paint contains heavy texturing material, it can settle in a matter of hours. Check the manufacturer's instructions to determine a paint's mixing requirements. Paints that have been recently delivered to a site may not require any mixing if they were mixed by the vendor.

When mixing paint, place the paddle into the container of paint before turning on the mixer. Start at the bottom of the container and run the mixer in a slow up-and-down motion to uniformly mix the solids with the liquids. To prevent paint splatters, always keep the paddles beneath the surface of the paint. Try to avoid overly vigorous mixing, which can lead to air entrapment in the paint. When this happens, let the paint sit for a few minutes to allow the air bubbles to rise to the top.

Mixers are used on compounds in which powders are mixed with water. They are also used whenever texturing material is added to a compound. Texturing material is usually perlite, vermiculite, or polystyrene. These materials are very light in comparison to the **viscosity** of the compound, so they won't settle out.

Many mixers are designed to minimize air entrapment in compounds. This is important because the compounds are thick, and air, which is light, has difficulty rising to the top. When a compound containing entrapped air is applied, the air bubbles may reach the surface before the compound dries. When the bubbles pop, craters or pits are formed on the surface of the compound. A few of these flaws are no problem, but a finish covered with them is unattractive and will need to be sanded and refinished.

When using a mixer, follow these guidelines:

Viscosity: The thickness of a fluid.

Step 1   Make sure that the mixer is clean before you use it. Dirt, dust, grease, or dried paint or compound can contaminate the paint or compound you are mixing.

Step 2   Put the mixer blade into the container before turning it on.

Step 3   When mixing powder compounds, sift the compound into clean water and mix it manually with a stir stick or with waterproof-gloved hands. Let the mixture sit for 10 to 15 minutes to thicken. Then add more water and mix it in with the mixer until the desired consistency is reached.

WARNING!

The dust from powder compounds can cause respiratory and eye irritations. Always use safety goggles and an N95 respirator when handling or working with these powder compounds.

Step 4 When mixing, move the blade slowly across the bottom of the container to thoroughly mix the contents.

Step 5 To prevent air entrapment when mixing compounds or paints, avoid moving the blade in a vigorous up and down motion.

Step 6 Thoroughly clean the mixing blade in a bucket of soapy water, or by spraying with a hose, to prevent material from drying on the blade.

3.3.0 Applying Primer

Primer is a paint that is applied to a bare surface to help prepare it for the final coat. Primer fills in small holes, blocks some stains to prevent them from bleeding through the final coat, and provides a better surface than drywall paper for adhering the top coat. Bare drywall is usually primed with a water-based primer. However, when texturing is applied, the primer will be oil-based. Oil-based primer protects the drywall from the water in the texturing compound. Surfaces primed with an oil-based primer can be painted with either a latex or an oil-based top coat. Surfaces primed with a latex primer may only be painted with a latex top coat.

Like other interior wall paint, primer can be applied with a sprayer or rolled on. The same rolling and spraying techniques used to apply a base coat or top coat of paint are used with primer.

Not all surfaces require primer, so check the project specifications to find out if one is required. If this is not specified, follow the manufacturer's instruction for the product being used as the finish coat. Contact the manufacturer if necessary.

Latex primer is usually applied at temperatures between 50°F and 90°F. Oil-based primers can be applied at temperatures as low as 40°F. Primers that are applied with a roller should not be thinned. Primers that are sprayed on are often thinned. Check the manufacturer's literature for instructions. Never thin primer more than directed by the manufacturer.

CAUTION

Paint contains a variety of substances that have the potential to cause allergic reactions in some people. Inhaling paint fumes can cause mild symptoms (itchy eyes, runny nose, sneezing, congestion) to more concerning symptoms (headaches, shortness of breath, wheezing, and coughing).

Some people experience an allergic reaction on the skin, known as *contact dermatitis*, after touching paint. Symptoms include rash, itchy skin, blisters, raised bumps (hives), swelling, pain, burning sensation, and blistering or flaking skin.

Always make sure your work area is well-ventilated, use respiratory protection if needed, and avoid direct contact with paint products. Seek medical attention for severe allergic reactions.

3.4.0 Spray Painting

Becoming skilled at applying paint with a sprayer requires practice. In addition, sprayers vary by type and by manufacturer. It is important to familiarize yourself with the tool you are using before applying it to a finished surface. First, study the operator's manual. Make sure that the machine is clean and in good working order before you load it with paint. Then practice on scrap pieces of drywall.

3.4.1 Calculating Paint Quantities

Before you begin painting, calculate the amount of paint needed to complete the job. Taking the time to accurately calculate the quantity of paint is a step that will prevent paint from being wasted. It will also prevent you from wasting time preparing or purchasing more paint.

Calculating Surface Area

To calculate the total quantity of paint needed for a room with the following wall lengths determine the total area of all surfaces to be painted. All of the walls listed here are 12' in height.

Wall A: Length = 34
Wall B: Length = 10
Wall C: Length = 7
Wall D: Length = 7
Wall E: Length = 10

Use the following steps to determine the total surface area:

Step 1 Measure the length and height of each wall, and then multiply these two numbers to determine the surface area. For the combined walls listed here, the surface area is determined as follows:

Wall A: Length × height = 34' × 12' = 408 ft²

Wall B: Length × height = 10' × 12' = 120 ft²

Wall C: Length × height = 7' × 12' = 84 ft²

Wall D: Length × height = 7' × 12' = 84 ft²

Wall E: Length × height = 10' × 12' = 120 ft²

Step 2 Add together the areas of all surfaces. For example, using the areas calculated in Step 1, the sum is as follows:

408 + 120 + 84 + 84 + 120 = 816 ft²

Step 3 Calculate the area for each opening in the walls. For example, assume that Wall A has a 10' by 4' door and two 6' by 4' windows. Their areas are calculated as follows:

Door = 10' × 4' = 40 ft²

Window One = 6' × 4' = 24 ft²

Window Two = 6' × 4' = 24 ft²

Step 4 Add the areas of all openings together. Using the areas calculated in Step 3, the sum is as follows:

40 + 24 + 24 = 88 ft²

Step 5 Subtract the sum of the openings from the surface area as follows:

816 − 88 = 728 ft²

Calculating Quantity of Paint Needed

Step 1 Divide the surface area by the manufacturer's stated coverage per gallon. One gallon of latex paint typically covers 400 to 450 square feet. Using 400 square feet per gallon, the calculation is as follows:

728 ft² ÷ 400 ft² per gallon = 1.8 gallons

Step 2 The minimum amount of paint you need to paint the room is 1.82 gallons. Because it is impossible to spray without wasting some paint, an allowance for waste should be added to this calculation. Allowing ten percent for waste, the calculation is as follows:

1.82 gallons × 1.10 percent = 2.002 gallons (usually rounded to two gallons)

NOTE

The calculations are shown for painting smooth finishes. For painting textured finishes, up to 50 percent more paint will be needed.

The paint may need to be thinned before it can be used in the sprayer. When calculating paint quantities, remember to account for the increased quantity of paint yielded by thinning. Manufacturers provide specific instructions for thinning each type of paint they make. For information about thinning the paint, check the operator's manual for the sprayer machine and the paint manufacturer's literature. Never thin the paint more than the manufacturer's recommendation.

3.4.2 Spraying Techniques and Tips

Follow these guidelines when using a spray painting machine:

- Adjust the machine pressure and gun orifice so that enough paint is released when the trigger is pulled.
- Hold the nozzle perpendicular to and 8 to 12 inches from the surface being painted. Start moving the gun across the test surface, and then pull the trigger to release the paint.
- Avoid moving your arm in a pronounced arc. This motion releases paint droplets into the air and will result in too little paint on the ends of the sweep stroke.
- Slowly sweep the gun back and forth over an area. Continue overlapping sweeps until the coating is uniformly applied.
- Adjust the speed of application and/or the paint flow until adequate coverage is achieved without any drips or runs.
- Release the trigger at the end of a sweep to avoid over-painting the ends.
- Adjust your equipment and application technique so that you can paint an area with a minimum number of sweeps.
- Clean up overspray and spills as soon as possible. Wet paint is easier to clean up than dry paint.

3.5.0 Applying Paint and Compound with Hand Tools

Texture brushes, stucco brushes, texture rollers, and trowels are all hand tools that are used for texturing, which is covered in the next section of this module. Although cutting-in can be time consuming, it is sometimes necessary to produce a clean, professional result. Similarly, the extra time and effort required to apply paint with a roller is warranted in certain situations to achieve the desired level of detail and coverage.

3.5.1 Cutting-in with an Angled Brush

When using an angled brush to paint surface edges, follow these steps to achieve clean, crisp edges:

Step 1   Hold the brush at the base of the bristles where they meet the handle (at the top of the metal band). Use a pencil-like grip with the thumb on the front, the index finger on the top side, and the three remaining fingers on the back.

Step 2   Pour a minimal amount of paint into a shallow container into which you can easily fit the brush while gripping it.

Step 3   Dip the paint at an angle into the container to wet only the top $1/3$ of the bristles. Use the side of the container to wipe the excess paint from both sides of the bristles.

Step 4   Place the brush just below the edge to be painted and make a downward stroke. This ensures that the brush is not overloaded with paint and makes the next stroke glide across the surface more easily.

Step 5   Carefully place the brush to one side of the edge to be painted, angle it so that the edge of the bristles is aligned exactly to the edge of the surface, and make a smooth horizontal—or vertical, depending on the direction of the edge—stroke (*Figure 13*).

Step 6   Gently drag the brush back over any areas to reduce the appearance of brush strokes.

Step 7   Cutting-in should produce a line of paint around the edges of the surface wide enough that the rest of the surface can be either rolled or sprayed without getting paint on the unpainted side of the edge, as in *Figure 14*.

Developing this technique takes practice. A few practice strokes on some scrap material may be helpful to determine the right angle and pressure to apply for the paint and surface with which you are working.

Figure 13 Cutting-in technique.
Source: ungvar/Shutterstock

Figure 14 Proper result of cutting-in before spraying.
Source: photovs/iStock

3.5.2 Applying Paint with a Roller

Rolling may be used to correct imperfections from spraying, to apply base and top coats in smaller areas, or to provide a rolled appearance in areas that call for a little more attention to detail. While there is some debate about which specific techniques produce superior results with a roller, experts generally agree on the following guidelines:

- Pour enough paint in the tray so that the well is full, but the paint stops at the bottom of the sloped portion of the tray. This will prevent paint from getting on the roller frame as you reload the roller cover. Pour spouts and tray liners can be used to prevent spills and reduce clean-up time.

- Roll the roller down the slope of the tray into the well and back up the slope again. Repeat this process until the roller cover is evenly coated.

- Do not oversaturate the roller cover. It should not be dripping.

- Apply the paint to the surface using long, smooth, diagonal strokes. Some recommend making an M shape, which is painted over with horizontal strokes. Others recommend a V or W shape. What is agreed upon is that the strokes should be long, slow, at least slightly diagonal (not straight up and down), and should overlap each other. Overlapping diagonal strokes cover the lines left behind by the roller's edge (*Figure 15*).

Figure 15 Rolling technique.
Source: Grigorev_Vladimir/Getty Images

- When using an extension pole, ensure it is screwed into the roller handle securely. Extend the pole the shortest distance possible to reach the top of the area you are painting.
- Roll back over each section while the paint is still wet to create a smooth, evenly coated finish.
- When the roller is no longer providing the same coverage using the same amount of pressure, it is time to reload it with paint.
- Each time the loader is reloaded, overlap your strokes on the edge of the area you just painted to blend them together.
- If a second coat is needed, wait until the first coat has dried.

CAUTION

Whether applying paint or primer with a sprayer or manually, it is essential to provide adequate ventilation in your work area. Open doors and windows, strategically place fans to push air out of your work space, and keep the area open even while taking breaks.

3.0.0 Section Review

1. Which of the following type of paint magnifies surface defects?
   a. Dark-colored, matte finish
   b. Dark-colored, high-gloss
   c. Light-colored, high-gloss
   d. Light-colored, matte finish

2. When sanding drywall, how can you avoid inhaling silica dust?
   a. Ventilate the area with a fan.
   b. Shake the dust out of your hair and clothing.
   c. Sand the area by hand.
   d. Wear respiratory protection and clean up with a vacuum.

3. True or False? When using a paint mixer, always turn on the mixer blade before putting it into the mixing container.
   a. True
   b. False

4. Which type of primer should be used on a surface prior to applying texturing compound to protect the drywall from the water in the compound?
   a. Water-based
   b. Latex
   c. Oil-based
   d. Double-coated

5. After dividing the total surface area to be painted by the manufacturer's stated coverage per gallon, what percent should be added to the total to account for waste (excluding textured paint)?
   a. 5
   b. 10
   c. 15
   d. 20

4.0.0 Interior Specialty Finishes

Performance Task

6. Apply, or explain the application of, specialty texture and faux finishes to an interior surface to achieve a desired appearance.

Objective

Identify the different types of interior specialty finishes.

a. Describe specialty texture finishes and the techniques used to apply them to interior surfaces.

b. Describe specialty faux finishes and the techniques used to apply them to interior surfaces.

Specialty finishes are used to make a space more visually interesting by adding complexity, dimension, and variety. They can be used to add a rustic feel, warmth, sophistication, or coziness, or otherwise alter the mood of a space.

4.1.0 Texturing

In addition to bringing visual appeal to a space, texturing may also be used to hide surface imperfections. Textured walls can also prevent a small amount of sound from being transferred between walls.

Texturing may be applied in one of two ways: by using a texturing compound on the surface before it is painted, or by using a textured paint. Texturing a surface takes a little more time than using a textured paint, but it will produce a more distinctive look.

Many joint compounds can be used as texturing material. Other texturing materials are specially designed for this purpose. Texturing may be sprayed onto a surface or applied by hand in a desired pattern. Some texturing materials are sprayed onto the surface and then finished by hand.

4.1.1 Texturing Compound

Texturing compounds are applied to a surface before it is painted. These compounds may be sprayed on, or they may be applied by hand and worked into the desired texture. Another option is to apply the compound with a sprayer and then work it into the desired texture by hand. Texturing compounds are similar to joint compounds, and some joint compounds may be used as texturing compounds. Texturing materials can come in powder form or as a ready-mixed product. Some compounds are smooth, while others contain some type of aggregate to give them more texture. The aggregate is usually perlite, vermiculite, polystyrene, or Styrofoam™. It may be fine, medium, or coarse in texture. Before starting any texturing job, it's a good idea to practice first on pieces of scrap drywall.

Texturing compound can be worked in numerous ways to achieve an attractive finish. For a light texture, the compound should be about $\frac{1}{8}$" thick. Deeper textures will require a thicker layer. Many textures are manually applied and worked into the final surface. Others are sprayed onto the surface and then manually worked to achieve the desired texture. Compound may be reworked with a roller, brush, sponge, rag, trowel, or blade until the desired finish is achieved.

Skip trowel is a texturing technique that is common on ceilings. A hawk and trowel, or sometimes a wide drywall knife, is used to unevenly apply a thin layer of drywall compound to the surface. The knife or trowel is dragged, or skipped, across the surface, leaving behind areas of uncoated surfaces. Unlike skim coating, the compound is thicker and left unsmoothed in an attractively rough pattern. *Figure 16* shows a skip trowel finish.

The light stippling effect shown in *Figure 17* is created by gently pressing a stippling brush into wet compound, or lightly rolling it with a stipple roller. Deeper stippling effects are created by pressing the same type of brush or roller more firmly into the compound. To create a swirl finish (*Figure 18*), a stiff-bristled brush is worked across the wet compound in a fanlike motion. Crow's foot patterns (*Figure 19*) are a classic texturing technique that can be created by rolling a textured roller over drywall mud or by pressing a soft brush with long bristles (known as a *stomp brush* or *slap brush*) or a stipple brush into wet mud, rotating the brush, and pressing again at an adjacent location on the surface. Another popular texture finish using drywall compound is known as *knock-down*, shown in *Figure 20*. Knock-down finishes are created by pressing a stipple brush into wet compound in random patterns and then using a knife or trowel to flatten the peaks created by stippling.

Figure 16 Skip trowel finish.
Source: Yuri Shebalius/Shutterstock

Figure 17 Light stippling.
Source: edwsg/Shutterstock

Figure 18 Swirl finish.
Source: LordRunar/iStock

Figure 19 Crow's foot.
Source: NUM LPPHOTO/Shutterstock

Figure 20 Knock-down finish.
Source: Alison Sanders/NCCER

4.1.2 Spray-On Textures

Texturing compounds can be sprayed onto a surface by machine and then left to dry so that the texture is raised. One such texture is called *popcorn* (*Figure 21*). The popularity of this style saw a drastic decline after the 1990s, but the texture is still used in some commercial applications due to its ease of application and modest ability to absorb sound. When it is used, it is most often on a ceiling because it is easily damaged.

Another popular texture is called *orange peel* (*Figure 22*). This texture is created by using the same type of compound used to create popcorn textures but thinned with water and sprayed at lower air pressure. This technique allows the globules of compound to spread a bit as they dry, creating an appearance somewhat like an orange peel. Orange peel texture can be applied in fine, medium, or heavy granules. The example in *Figure 22* is fine to medium graininess.

Knock-down texture, as shown in *Figure 20*, can also be sprayed on a surface using the same compound as orange peel, with or without the aggregate.

Figure 21 Popcorn texture.
Source: h.yegho/Shutterstock

Figure 22 Orange peel texture.
Source: Sam Supachat/Shutterstock

Follow these guidelines when applying texturing compounds with a sprayer:

- Spraying can be very messy. Be sure to dress for the job. Remember to take precautions to prevent yourself from inhaling dust from powdered compound mixtures, such as wearing an N95 respirator, cleaning your clothing, and washing your hands.
- Mask off or cover anything that should not be sprayed.
- When applying knock-down texture, timing is important. Adjust your speed so that the sprayed area can be worked before it becomes too dry. Depending on the surface area being covered, it may be necessary to work with a partner, with one person spraying while the other applies the texturing effects.
- Clean up overspray promptly. To remove dried compound, spray it with water, allow it to sit for a minute or two, and then scrape it with a putty knife.

4.1.3 Textured Paints

Unlike texturing compound, which contains only the drywall compound and texturing material, textured paint is paint in which textured material has already been added. Textured paints eliminate the need to apply a separate coat of paint, as would need to be done over texturing compound. Textured paints are applied in the same manner as smooth paints, but with slight differences in technique due to the extra thickness of textured paint. Most may be rolled, brushed, or sprayed on. Some types of textured paints require periodic mixing during application to redistribute the texturing material. Check the manufacturer's instructions for details.

Textured paints are thicker than regular paint. When they are applied to a surface, they can be manually worked to achieve a certain look. Texture granules mixed into paint can be fine, medium, or coarse. Texturing material may be in the base coat, which is usually white or off-white. It may also be mixed into the paint that will be used for the final coat. Textured paints are similar to textured surfaces in their ability to hide surface flaws. These paints may be sprayed, brushed, or rolled onto a surface in accordance with the manufacturer's instructions.

Pre-mixed textured paints can be used to create finishes that resemble those created with drywall compound, such as orange peel, knockdown, and popcorn. Another option is to mix texture additives into the paint before application. Stone and sand (*Figure 23*) textures are popular additives.

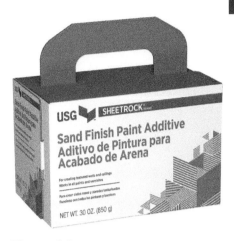

Figure 23 Sand texture additive.
Source: Courtesy of USG

CAUTION

Some paints and solvents are classified as hazardous waste. These products are dangerous or potentially dangerous to human health or the environment. Hazardous waste must be disposed of in accordance with the Environmental Protection Agency (EPA) Resource Conservation and Recovery Act (RCRA) *40 CFR 261*, or with local law. Local law can be stricter than the RCRA, but it cannot be more lenient. The Safety Data Sheet (SDS) will tell you how to dispose of a product.

If a hazardous material gets spilled in an area, check the SDS for cleanup instructions. The EPA may need to be informed of the spill, depending on the material and the size of the area affected. Contact your local and statewide environmental protection agencies to make sure you are informed in the event that a spill occurs.

4.2.0   Faux Finishes

Faux, pronounced like the English word *foe*, is a French word meaning fake or imitation. Faux finishes imitate other styles or naturally occurring materials. Faux finishes use paint to make a surface resemble a variety of natural materials like marble, granite, and wood grain. Faux finishes can also be applied to mimic distressed, weathered, cracked, sun-bleached, and other aged surfaces. Some faux finishes imitate crafted finishes such as stucco, limewashing, and fabric wall coverings.

Faux finishes are made by applying layers of paint on a surface. Applying faux finishes is time intensive because of the attention to detail required, so they are often used in a limited area. Two popular faux finishes are marbling and Venetian plaster. Marbling uses a number of layers of paint that are applied in a particular way to give the final surface the appearance of marble. This finish is often used on columns. Venetian plaster simulates a polished plaster finish popular in Europe. Faux finishes should look natural and have random patterns with varying color shades. Applying them well is a skill that takes practice. Before using these faux finish techniques on the job, it's a good idea to perfect your skills on scraps of drywall.

4.2.1 Marble Faux Finish

Applying a marble faux finish (*Figure 24*) to a surface takes time and patience. For this finish, you will need the following supplies:

- Latex satin paint (one base color and at least two other shades)
- Latex satin paint for veining (white, gray, green, or another shade of the base color)
- Clear polyurethane varnish or wax
- Paint roller and holder and paint pan
- Natural sea sponge
- Pointed artist's brush or feather

Figure 24 Faux marble finish.
Source: Sergey Scherbak/Shutterstock

Natural marble contains many shades of color. The more shades of color you use, the more realistic the finish will look, but use the paint sparingly. The following is a general procedure for applying a marbled faux finish. As you gain experience, you will likely develop your own technique.

Step 1   Apply the base coat over the entire surface with a damp sponge, roller, or sprayer. This coat does not need to cover the surface completely so that none of the surface color shows through. When rolling the paint, be sure to smooth out all roller marks. Allow this coat to dry.

Step 2   Alternate applying the two lighter colors with a damp sponge until the desired effect is obtained. Using a light touch, apply the sponge directly to the surface, and then remove it. Be careful not to drag the sponge across the surface. Some drywall technicians allow this coat to dry before moving on to the next step. Others proceed immediately to the next step.

Step 3   Using an artist's brush and the veining paint, lightly paint veins on the surface. Be sure to vary the appearance and size of the veins. The key to this technique is to use a light hand. Veins should be delicate. Allow this coat to dry.

Step 4   To seal the coating, apply a faux finish varnish or wax over the surface according to the manufacturer's instructions.

4.2.2 Venetian Plaster

Creating a Venetian plaster faux finish (*Figure 25*) starts with Venetian plaster or Venetian plaster paint. To achieve this faux finish, two or three layers of plaster or paint are applied to a surface with a trowel. After the paint or plaster has dried, the surface is polished. This polishing, also called *burnishing*, is what gives the plaster its glossy finish.

Venetian plaster paint is thicker than regular latex paint. It is specially formulated to achieve the desired surface. Be sure to stir the paint thoroughly before use. Also, wipe the application blade from time to time to avoid contaminating the surface with dried plaster. Practice on scrap material until you achieve the desired result. To create this faux finish, you will need the following supplies:

• Venetian plaster or Venetian plaster paint

• Venetian plaster or paint top coat

• Steel trowel or blade with rounded corners

• Clean cloth, power buffer, or power sander with 600-grit sandpaper

Figure 25 Venetian plaster faux finish in progress.
Source: Guillem Lopez Borras/Shutterstok

The following is a general procedure for applying a Venetian plaster faux finish. Be sure to follow the manufacturer's application instructions for the product being used.

Step 1 Apply plaster or paint to the surface with a trowel or blade. Hold the tool so that the blade is at a 15- to 45-degree angle to the surface. Spread the plaster or paint thinly with a rough finish. Apply the coating in a random X-pattern using long and short strokes. Smooth out any obvious high spots but leave the surface rough. This coat does not need to cover the surface completely, because another coat will be applied. Allow the first coat to dry.

Step 2 Apply the second layer of plaster in a manner similar to the first but hold the tool so the blade is at a 45- to 90-degree angle to the surface. Make sure that this coat completely covers the surface. Allow the second coat to dry.

Step 3 If desired, apply a third layer of plaster, as you did with the second coat, and allow this coat to dry.

Step 4 Using a moistened rag, rub the surface of the plaster until a glossy sheen appears. Instead of a rag, you can use a power buffer or sander to lightly polish the plaster.

Step 5 Apply the protective topcoat to the plaster.

NOTE

Some products recommend applying the top coat to the plaster before it is polished.

4.0.0 Section Review

1. Which technique is used to hide surface imperfections, prevent a small amount of sound from being transferred, and/or bring visual appeal to a space?
   a. Faux painting
   b. Manual painting
   c. Texturing
   d. Spray painting

2. At which point in the painting process are texturing compounds applied to the surface?
   a. During
   b. Before
   c. After
   d. After the paint is dry

3. What should be applied to dried texturing compound so that it can be removed if mistakes are made during the application process?
   a. Solvent
   b. All-purpose cleaner
   c. Water
   d. Soapy water

4. What documentation should you check for cleanup instructions if a hazardous material is spilled?
   a. MSDS
   b. Material's container
   c. Manufacturer's website
   d. SDS

5. Which time-intensive technique is used to make a surface resemble the appearance of naturally occurring materials like marble, granite, and wood grain?
   a. Spray painting
   b. Faux finishes
   c. Texturing compound
   d. Cutting-in

1. The paint component that determines the finish color is the _____.
   a. Additive
   b. Pigment
   c. Binder
   d. Solvent

2. Latex, or water-based paint _____.
   a. takes a long time to dry
   b. is available in a limited number of colors
   c. has low odor
   d. has poor coverage

3. The potlife for epoxy paints after they are mixed is _____.
   a. a few hours
   b. several days
   c. a month or more
   d. usually more than three days

4. Which of the following is an appropriate solvent to clean a paint sprayer after it has been used to apply latex paint?
   a. Water
   b. Bleach
   c. Detergent
   d. Turpentine

5. Which brush is ideal for most edging (cutting-in) tasks because it is wide enough to paint away from the protected surface but small enough to maneuver easily?
   a. 2"–3", straight
   b. 2"–3", angled
   c. 4"–5", stiff-bristled
   d. 3"–4", synthetic

6. Which of the following is true regarding surface preparation of drywall before painting?
   a. Inspecting the surface is not necessary.
   b. Always inspect the surface using a systematic approach.
   c. Inspect for major problems and repair smaller issues as you paint.
   d. Inspection is not the responsibility of the painter.

7. Which tool should be used first to remove drywall dust from a surface before painting?
   a. A vacuum with a soft-bristled head
   b. A wet sponge
   c. A damp cloth
   d. A large, flat mop

8. To mix powder compounds, _____.
   a. always use a mixer
   b. initially sift the powder into the water and mix by hand
   c. add enough powder to achieve the desired consistency
   d. sift the powder into a clean container and then add clean water

9. True or False? You should always prime a drywall surface before applying a finish coat.
   a. True
   b. False

10. Latex primers should be applied at temperatures between _____.
    a. 30°F and 60°F
    b. 50°F and 90°F
    c. 70°F and 90°F
    d. 60°F and 80°F

11. When painting with a sprayer, adjust your equipment so that you can paint the area with _____.
    a. an arc-like motion
    b. a minimum number of sweeps
    c. as much paint as possible
    d. no protective gear

12. Texturing may be applied by using a texturing compound on the surface before painting, or by using a textured _____.
    a. paint
    b. epoxy
    c. enamel
    d. primer

13. The aggregate in a texturing compound may be _____.
    a. latex
    b. acrylic
    c. powder
    d. coarse, medium, or fine

14. To achieve an orange peel textured finish, the same compound as popcorn textured finish may be used but _____.
    a. thinned with water and sprayed at lower air pressure
    b. manually applied
    c. sprayed with water after it is applied
    d. thinned and applied with a roller or brush

15. What is another term for the polishing of Venetian plaster faux finishes after the paint or plaster has dried, resulting in a glossy appearance?
    a. Burnishing
    b. Shining
    c. Waxing
    d. Buffing

Answers to odd-numbered questions are found in the Module Review Answer Key at the back of this book.

Answers to Section Review Questions

Answer	Section	Objective
Section One		
1. c	1.1.1	1a
2. b	1.1.4	1a
3. d	1.2.1	1b
4. d	1.2.5; *Table 1*	1b
Section Two		
1. c	2.1.0	2a
2. a	2.2.0	2b
3. b	2.3.1	2c
4. d	2.3.3; *Table 3*	2c
Section Three		
1. b	3.1.0	3a
2. d	3.1.0	3a
3. b	3.2.0	3b
4. c	3.3.0	3c
5. b	3.4.1	3d
Section Four		
1. c	4.1.0	4a
2. b	4.1.1	4a
3. c	4.1.2	4a
4. d	4.1.3	4a
5. b	4.2.0	4b

APPENDIX 45202A

Common Terms Used in Cold-Rolled Steel Framing Work

AISC: American Institute of Steel Construction.

AISI: American Iron and Steel Institute.

Applicable building code: Building code under which the building is designed.

Approved: Approved by a building official or design professional.

Base steel thickness: The thickness of bare steel exclusive of all coatings.

Bracing: Structural elements that are installed to provide restraint or support (or both) to other framing members so that the complete assembly forms a stable structure.

Ceiling joist: A horizontal structural framing member that supports ceiling components and which may be subject to attic loads.

Cold-formed sheet steel: Sheet steel or strip steel made from relatively thin metallic-coated coils, formed by processes carried out at or near ambient room temperature. This is different from hot-rolled steel, where uncoated steel's temperature is raised to near its melting point for forming into final shapes. Cold-forming processes typically include slitting, shearing, press braking, and roll forming.

Cold-formed steel: See *cold-formed sheet steel*.

Cold-formed steel structural framing: The elements of the structural frame, as given in the Code of Standard Practice, Section B1.

Component assembly: A fabricated assembly of cold-formed steel structural members that is made by the component manufacturer; may also include structural steel framing, sheathing, insulation, or other products.

Component design drawing: The written, graphic and pictorial definition of an individual component assembly, which includes engineering design data.

Component designer: The individual or organization responsible for the engineering design of component assemblies.

Component manufacturer: The individual or organization in charge of the manufacturing of component assemblies for the project.

Component placement diagram: The illustration, supplied by the component manufacturer, identifying the assumed location for each of the component assemblies and referencing each individually designated component design drawing.

Construction manager: The individual or organization selected by the owner to issue contracts for the construction of the project and to purchase products.

Contract documents: The documents including, but not limited to, plans and specifications that define the responsibilities of the parties involved in bidding, purchasing, designing, supplying, and installing cold-formed steel framing.

Contractor: The individual or organization that is contracted with to assume full responsibility for the construction of the structure.

Cripple stud: A stud that is placed between a header and a window or door head track, a header and wall top track, or a windowsill and a bottom track to provide a backing to attach finishing and sheathing material.

Design professional: A person who is registered or licensed to practice a design profession as defined by the statutory requirements of the state in which the project is to be constructed.

Design thickness: The steel thickness used in design which is equal to the minimum base steel thickness divided by 0.95.

Drawings: See plans and installation drawings.

Edge stiffener: That part of a C-shape framing member that extends from the flange as a stiffening element that extends perpendicular to the flange.

Erection drawings: See *installation drawings*.

Erector: See *installer*.

Flange: That portion of the C-shape framing member or track that is perpendicular to the web.

Floor joist: A horizontal structural framing member that supports floor loads and superimposed vertical loads.

Framing contractor: See *installer*.

Framing material: Steel products, including but not limited to, structural members and prefabricated structural assemblies, ordered expressly for the requirements of the project.

General contractor: See *installer*.

Harsh environments: Coastal areas where additional corrosion protection may be necessary.

In-line framing: Framing method in which all vertical and horizontal load-carrying members are aligned.

Installation drawings: Field illustrations that show the location and placement of the cold-formed steel structural framing.

Installer: Party responsible for the installation of cold-formed steel products.

Jack stud: A stud that does not span the full height of the wall and provides bearing for headers. Also called a *trimmer stud*.

King stud: A stud, adjacent to a jack stud, which spans the full height of the wall and supports vertical and lateral loads.

Lip: See *edge stiffener*.

Material supplier: An individual or entity responsible for furnishing framing materials for the project.

Nonstructural member: A member in a steel-framed assembly which is limited to a transverse load of not more than 10 lb/ft^2 (480 Pa); a superimposed axial load, exclusive of sheathing materials, of not more than 100 lb/ft (1,460 N/m); or a superimposed axial load of not more than 200 lbs (890 N).

Owner: The individual or entity organizing and financing the design and construction of the project.

Plans: Drawings prepared by the design professional for the owner of the project. The drawings include, but are not limited to, floor plans, framing plans, elevations, sections, details, and schedules as necessary to define the desired construction.

Punchout: A hole made in the web of a steel framing member during the manufacturing process.

Shop drawings: Illustrations for the production of individual component assemblies for the project.

Span: The clear, horizontal distance between bearing supports.

Specifications: Written instructions that, with the plans, define the materials, standards, product designs, and workmanship expected on a construction project.

Standard cold-formed steel structural shapes: Cold-formed steel structural members that meet the requirements of the SSMA Product Technical Information.

Strap: Flat or coil sheet steel material, typically used for bracing and blocking, that transfers loads by tension and/or shear.

Structural engineer-of-record: The design professional responsible for sealing the contract documents indicating that he or she has performed or supervised the analysis, design, and document preparation for the structure and has knowledge of the requirements for the load-carrying structural system.

Structural member: A member that is designed or intended to carry loads, such as a floor joist, rim track, structural stud, wall track in a structural wall, ceiling joist, roof rafter, or header.

Structural stud: A stud in an exterior wall, or an interior stud that supports superimposed vertical loads and may transfer lateral loads, including full-height wall studs, king studs, jack studs, and cripple studs.

Stud: A vertical framing member in a wall system or assembly.

Trimmer: See *jack stud*.

Truss: A coplanar system of structural members joined together at their ends, usually to construct a series of triangles that form a stable, beamlike framework.

Yield strength: A characteristic of the basic strength of the steel material defined as the highest unit stress that the material can endure before permanent deformation occurs as measured by a tensile test in accordance with ASTM A370.

Module Review Answer Key

MODULE 01 (45201)

Answer	Section Head
1. b	1.1.0
3. d	1.1.3
5. a	1.1.4
7. b	2.1.1
9. c	2.1.2
11. d	2.1.4
13. b	2.1.5
15. a	2.2.1
17. c	2.2.1
19. c	2.2.2
21. d	2.3.1
23. a	2.4.0

MODULE 02 (45202)

Answer	Section Head
1. d	1.2.1
3. c	1.2.1
5. b	1.2.2
7. c	1.2.5
9. b	2.1.0
11. d	2.1.1
13. c	2.2.1
15. a	2.2.4
17. d	2.4.1
19. d	2.4.2

MODULE 03 (27209)

Answer	Section Head
1. b	1.0.0
3. a	1.0.0
5. b	1.1.3
7. d	1.2.1
9. c	1.2.1
11. a	1.3.2
13. d	2.1.0
15. c	3.2.0
17. c	3.2.2
19. b	3.4.2

MODULE 04 (45204)

Answer	Section Head
1. b	1.1.1
3. b	1.1.3
5. d	1.1.4
7. c	1.2.0
9. b	1.3.0
11. a	1.4.0
13. d	1.5.0
15. b	2.1.1
17. a	2.1.1

MODULE 05 (45205)

Answer	Section Head
1. b	1.0.0
3. d	2.1.1
5. b	2.1.3
7. b	3.1.0
9. a	3.1.0
11. b	3.1.0
13. c	3.2.0
15. c	3.2.0
17. c	3.2.0
19. a	3.3.0
21. c	3.4.0
23. c	3.4.0

MODULE 06 (45206)

Answer	Section Head
1. b	1.1.1
3. a	1.2.4
5. b	2.3.3
7. a	3.1.0
9. b	3.3.0
11. b	3.4.2
13. d	4.1.1
15. a	4.2.2

GLOSSARY

Acoustical materials: Types of ceiling panel, plaster, and other materials that have high absorption characteristics for sound waves.

Acoustics: A science involving the production, transmission, reception, and effects of sound. In a room or other location, it refers to those characteristics that control reflections of sound waves and thus the sound reception in the area.

Acrylic latex: A type of paint that is water-based with an acrylic binder. Acrylic latex paints have the enduring color, cleanability, and adhesion of latex with the added durability of acrylic resin.

ADA Requirements: The Americans with Disabilities Act requires the use of accessible design methods for the construction of public accommodations such as lodging, public parks, schools, and restaurants.

As-built: Revised drawing of an installation that shows the changes made during installation.

A-weighted decibel (dBA): A single number measurement based on the decibel but weighted to approximate the response of the human ear with respect to frequencies.

Beams: Load-bearing horizontal framing elements supported by walls or columns and girders.

Blocking: C-shaped track, break shape, or flat strap material attached to structural members, flat strap, or sheathing panels to transfer shear forces.

Building codes: Codes published by state and local governments to establish minimum standards for various types of interior and exterior construction.

Callouts: Markings or identifying tags describing parts of a drawing; callouts may refer to detail drawings, schedules, or other drawings.

Ceiling panels: Acoustical ceiling boards that are suspended by a concealed grid mounting system. The edges are often kerfed and cut back.

Ceiling tiles: Any lay-in acoustical boards designed for use with exposed grid mounting systems. Ceiling tiles normally do not have finished edges or precise dimensional tolerances because the exposed grid mounting system provides the trim-out.

Civil drawings: Drawings that show the overall shape of the building site. They are also called *site plans*.

Clip angles: L-shaped pieces of steel (normally having a 90-degree bend), typically used for connections.

Cold-formed steel: Steel products made from relatively thin metallic-coated coils, formed by processes carried out at or near ambient room temperature. These processes are different from hot-rolled steel, where uncoated steel's temperature is raised to near its melting point for forming into final shapes. Cold-forming processes typically include slitting, shearing, press braking, and roll forming.

Compressive strength: Refers to a material's ability to support weight.

Contour lines: Imaginary lines on a site plan/plot plan that connect points of the same elevation. Contour lines never cross each other.

C-shape: A cold-formed steel shape used for structural and nonstructural framing members consisting of a web, two flanges, and two lips (edge stiffeners).

Curtain wall: A light, nonbearing exterior wall attached to the concrete or steel structure of a building. It primarily resists wind loads and supports only its own weight and cladding weight.

Cut-in: A technique used to manually paint around a fixture with a brush or roller.

Decibel (dB): An expression of the relative loudness of sounds in air as perceived by the human ear.

Diaphragms: Floor, ceiling, or roof assemblies designed to resist in-plane forces such as wind or seismic loads.

Diffuser: An attachment for duct openings in air distribution systems that distributes the air in wide flow patterns. In lighting systems, it is an attachment used to redirect or scatter the light from a light source.

Dry lines: A string line suspended from two points and used as a guideline when installing a suspended ceiling.

Elevation views: Drawings giving a view from the front, rear, or side of a structure.

Fissured: A ceiling-panel or ceiling-tile surface design that has the appearance of splits or cracks.

Flashing: Thin strips of material that are placed around openings in a structure to divert water away from the opening.

For-fill: A wall to be filled with some type of material, such as concrete or insulation.

Frequency: Cycles per unit of time, usually expressed in hertz (Hz).

Girders: Large steel or wooden beams supporting a building, usually around the perimeter.

Headers: Horizontal structural framing members used over floor, roof, or wall openings to transfer loads around the opening to supporting structural framing members.

Hertz (Hz): A unit of frequency equal to one cycle per second.

Isometric drawing: A three-dimensional type of drawing in which the object is tilted so that all three faces are equally inclined to the picture plane.

Jambs: The top and sides of a door or window frame that are in contact with the door or sash.

Joists: Horizontal members of wood or steel supported by beams and holding up the planks of floors or the laths of ceilings. Joists are laid edgewise to form the floor support.

Knurled: A series of small ridges, dimples, or embossments used to provide a better gripping surface on metal.

Landscape drawings: Drawings that show proposed plantings and other landscape features.

Lateral: Running side to side; horizontal.

Lips: That part of a C-shape framing member that extends from the flange as a stiffening element that extends perpendicular to the flange; also called *edge stiffene*.

Mils: Units of measurement equal to $1/1,000$ of an inch.

Orifice: The opening in the nozzle of a paint sprayer machine through which paint is sprayed.

Panelization: The process of assembling steel-framed walls, joists, or trusses before they are installed in a structure. Roll-formers normally cut studs to within 1/8-inch tolerance. This helps the framer to consistently create straight walls on the panel table that are easy to install in the field.

Permeable: Having pores or openings that permit liquids, such as water, or gases to pass through.

Pigment-volume concentration (PVC): The relationship between the amount of pigment and the amount of binder in paint. Higher PVC levels indicate more pigment compared to binder.

Plan view: A drawing that represents a view looking down on an object.

Plenum: A chamber or container for moving air under a slight pressure. In commercial construction, the area between the suspended ceiling and the floor or roof above is often used as the HVAC return air plenum.

Plenum: A confined space, such as the area between a suspended ceiling and an overhead deck, that is used as a return for a heating, cooling, and ventilation system.

Potlife: The period of time in which a mixed, two-component epoxy paint must be used.

Racking: Being forced out of plumb by wind or seismic forces.

Request for information (RFI): A document used during the construction process that is used to clarify and resolve questions and information gaps in plans, drawings, specifications, and agreements.

Rim track: A horizontal structural member that is connected to the end of a floor joist.

Riser diagrams: Isometric drawings that depict the layout, components, and connections of a piping system.

Schedules: Charts or tables that provide detailed information corresponding to various parts of the drawings.

Sections: Drawings that detail construction techniques or materials.

Shear wall: A wall designed to resist lateral forces such as those caused by earthquakes or wind.

Spandrel beams: Structural support elements at the outer edge of a building or edge of a slab.

Striated: A ceiling-panel or ceiling-tile surface design that has the appearance of fine parallel grooves.

Tensile strength: Refers to the flexibility of a material.

Title block: A section of an engineering drawing blocked off for pertinent information, such as the title, drawing number, date, scale, material, draftsperson, and tolerances.

Tracks: Framing members consisting of only a web and two flanges. Track web depth measurements are taken to the inside of the flanges.

Veneer: A thin sheet of material, such as wood, that is glued to some type of fabricated material.

Viscosity: The thickness of a fluid.

REFERENCES

45201 Commercial Drawings

Architectural Graphic Standards. Twelfth Edition, 2016. The American Institute of Architects. Hoboken, NJ: John Wiley & Sons.

Digital Codes. International Code Council. *https://codes.iccsafe.org/*

Understanding Construction Drawings. Seventh Edition, 2018. Boston, MA: Cengage Learning.

45202 Steel Framing

American Iron and Steel Institute (AISI): *www.steel.org*

BuildSteel: *www.buildsteel.org*

Cold-Formed Steel Engineers Institute (CFSEI): *www.cfsei.org*

Steel Framing Industry Association (SFIA): *www.cfsteel.org*

Steel Stud Manufacturers Association (SSMA): *www.ssma.com*

27209 Suspended and Acoustical Ceilings

Ceiling and Interior Systems Construction Association, *https://www.cisca.org*.

International Building Code®. 2021. International Code Council.

International Residential Code®. 2021. International Code Council.

National Earthquake Hazards Reduction Program, FEMA, *https://www.nehrp.gov/*.

The Gypsum Construction Handbook. 7th edition. 2014. Chicago: USG.

US Geological Survey, *https://www.usgs.gov/*.

45204 Interior Specialties

Acoustics and Sound Insulation: Principles, Planning, Examples, 2009. Eckard Mommertz. Munich, Germany: Institut fuer Internationale Architektur-Dokumentation.

Application and Finishing of Gypsum Panel Products: GA-216-2021, 2021. Gypsum Association. Orland Park, IL: American Technical Publishers.

The Gypsum Construction Handbook, 2014. USG. Greenville, SC: RSMeans.

45204 Exterior Cladding

http://www.cement.org

http://www.culturedstone.com (1-800-255-1727)

http://www.eima.com

http://www.jameshardiecommercial.com

https://stuccomfgassoc.com

45206 Interior Finishes

Drywall Finishing Glossary: Key Terms to Know. 2023. National Gypsum. *https://www.nationalgypsum.com/who-we-are/blog/building-knowledge/drywall-finishing-terms-to-know*

2023 National Painting Cost Estimator. 2023. Dennis D. Gleason, CPE. Carlsbad, CA: Craftsman Book Company.